SpringerBriefs in Electrical and Computer Engineering

T0214553

For further volumes:
http://www.springer.com/series/10059

Panagiotis Symeonidis • Dimitrios Ntempos
Yannis Manolopoulos

Recommender Systems for Location-based Social Networks

Springer

Panagiotis Symeonidis
Department of Informatics
  Data Engineering Laboratory
Aristotle University of Thessaloniki
Stavroupoli, Thessaloniki
Greece

Yannis Manolopoulos
Department of Informatics
  Data Engineering Lab
Aristotle University of Thessaloniki
Stavroupoli, Thessaloniki
Greece

Dimitrios Ntempos
Kiwe Development
Kalamaria, Thessaloniki
Greece

ISSN 2191-8112   ISSN 2191-8120 (electronic)
ISBN 978-1-4939-0285-9   ISBN 978-1-4939-0286-6 (eBook)
DOI 10.1007/978-1-4939-0286-6
Springer New York Heidelberg Dordrecht London

Library of Congress Control Number: 2014930156

Printed on acid-free paper

Springer is part of Springer Science+Business Media (www.springer.com)

Contents

Chapter 1
Introduction

Social networking sites, such as Facebook and LinkedIn, have attracted a huge attention after the widespread adoption of Web 2.0 technology. These systems contain data, which can be mined and used for making personalized predictions and recommendations of products, users and digital content. In particular, they collect information from users' social contacts and their interactions (co-tagging of photos, co-rating of products etc.) and make recommendations of products or even people to users based on their common friends, common commenting on written posts etc.

Mobile communication devices reach now every corner of planet earth. People are inextricably linked to the internet through mobile devices that allow ubiquitous and pervasive computing. Lately, technological progressions in mobile devices (GPS, Wi-Fi) enabled the incorporation of geo-location data in the traditional web-based online social networks, which have been evolved to Location-based Social Networks (LBSNs). Geo-location data seems to serve as the physical dimension that Web lacks. In this new era, users can benefit by getting pervasive and ubiquitous access to location-based services from anywhere via mobile devices. Moreover, users can share location-related information with each other to leverage their collaborative social knowledge.

The goal of this book is to bring together important research in a new family of recommender systems aimed at serving LBSNs. The chapters introduce a wide variety of recent approaches, from the most basic to the state-of-the-art, for providing recommendations in LBSNs. The material covered in the book is addressed to graduate students, teachers, researchers, and practitioners in the areas of web data mining, information retrieval, and machine learning. The book is organized into three parts. Part I provides introductory material on recommender systems, online social networks and LBSNs. Part presents a wide variety of recommendation algorithms, ranging from the most basic methods to the state-of-the-art, as well as a comparison of the characteristics of these recommender systems. Part III provides a step-by-step case study on the technical aspects of deploying and evaluating a real-world LBSN, which provides location, activity and friend recommendations.

In the sequel, a brief introduction to each chapter of the book follows:

P. Symeonidis et al., *Recommender Systems for Location-based Social Networks*,
SpringerBriefs in Electrical and Computer Engineering,
DOI 10.1007/978-1-4939-0286-6_1, © The Author(s) 2014

Chapter 2: Recommender Systems

Recommender systems base their operation on past user purchases/ratings over a collection of items, for instance, books, CDs, etc. Collaborative Filtering (CF) is a successful recommendation technique that confronts the "information overload" problem. Memory-based algorithms recommend according to the preferences of nearest neighbors, and model-based algorithms recommend by first developing a model of user ratings. In this chapter, we bring to surface factors that affect recommendation process. Moreover, we describe the most important problems related to recommender systems and give some references to actual solutions. Finally, there is an economic and social report regarding recommender systems, which examines them under a rather market-based angle.

Chapter 3: Online Social Networks

This chapter provides: (1) some definitions and basic concepts for Online Social Networks (OSNs), (2) a brief literature review of OSNs, (3) some paradigms of commercial OSNs, and (4) the transition of OSNs towards location as an auxiliary dimension. Finally, the social and economic report of commercial OSNs helps the reader to realize the huge potential that Location-based Social Networks (LBSNs) have, based on the fact that OSNs have incorporated the location dimension in recent years.

Chapter 4: Location-Based Social Networks

Location-based Social Networks (LBSNs) can be considered as a special OSN category. Actually, an LBSN has the same OSN's properties, but considers location as the core object of its structure. This chapter initially provides some definitions and basic services that are offered by LBSNs, a brief literature review, and two commercial paradigms of LBSNs. Additionally, a few location-based research projects are presented. Moreover, there is an economic and social report regarding LBSNs, which aims to investigate the field under a different, more market oriented prism. The last section provides an example of how a recommender system can benefit an LBSN.

Chapter 5: Framework

This chapter introduces the challenges that recommendation algorithms have to overcome in LBSNs. We also present the main algorithmic categories in the field of LBSNs (i.e. Collaborative Filtering, Semantically-enhanced, etc.). Moreover, we introduce the four types of recommendations in LBSNs (i.e. location, activity, friend, event). Finally, the reader meets an experimental framework for evaluating the quality of recommendations in LBSNs.

Chapter 6: Algorithms

This chapter provides more details on advanced research work proposed in LBSNs, and deepens in the algorithmic side of each method. We present algorithms for generic and personalized recommendations. For readability reasons, we have categorized the state-of-the-art methods in different algorithmic families such as matrix and tensor factorization, graph-based methods, and hybrid models.

Chapter 7: Comparison

This chapter compares and categorizes the algorithms that are described in Chap. 6 based on their basic characteristics. We categorized them based on (1) the kind of recommendation they provide (i.e., generic or personalized), (2) the type of recommendation they provide (i.e. Friend, Location, Activity, and Event), (3) the data representation they use for their model (i.e. matrix, tensor, graph), (4) the technique they are based on (i.e. probabilistic, semantic, collaborative filtering, etc.), (5) the data sets and the metrics they use in their experiments. The aforementioned categorizations help the reader to understand the main research choices that have been proposed in the research field of LBSNs and provides insight for further directions in the future.

Chapter 8: Real World Location-Based Social Network

This chapter presents a real-world recommender system for LBSNs. GeoSocialRec allows to test, evaluate and compare different recommendation styles in an online setting, where the users of GeoSocialRec actually receive recommendations during their check-in process. The system's experimental evaluation checks its performance in terms of accurate recommendations. Moreover, we present a user study for evaluating different styles of explanations that come along with a recommendation to users.

Chapter 9: Conclusion

The last chapter concludes the book with a summary and future research directions.

Part I
Basic Definitions and Concepts

Chapter 2
Recommender Systems

Recommender systems base their operation on past user purchases/ratings over a collection of items, for instance, books, CDs, etc. In this chapter, we bring to surface factors that affect the recommendation algorithmic process. Moreover, we describe the most important problems related to recommender systems and give some references to actual solutions. Finally, there is an economic and social report regarding Recommender Systems, which examines them under a more market-based angle.

2.1 Basic Approaches

Recommender systems are gaining widespread acceptance in e-commerce applications as a way of tackling the "information overload" problem. This problem affects our everyday experience while searching for information on a topic. To overcome this problem, we often rely on suggestions from others who have more experience on the topic. However, in the Web case where there are numerous suggestions, it is not easy to detect the trustworthy ones. The process of recommendation becomes controllable by shifting from individual to collective suggestions,

Three parallel approaches have emerged in the context of recommender systems: collaborative filtering (CF), content-based Filtering (CB) and hybrid methods.

Collaborative filtering algorithms recommend those items to the target user, that have been rated highly by other users with similar preferences and tastes [24, 28]. In most CF approaches, only the item and users' identifiers are accessible and no additional information over items or users is provided. Websites that provide recommendations in the form, "Customers who bought item i also bought item y", typically fall under collaborative filtering approaches. Grouplens research group [24] introduced a collaborative filtering algorithm, known as user-based CF, because it employs users' similarities for the formation of the neighborhood of nearest users. Another CF algorithm proposed by Sarwar et al. [28], is known as

P. Symeonidis et al., *Recommender Systems for Location-based Social Networks*,
SpringerBriefs in Electrical and Computer Engineering,
DOI 10.1007/978-1-4939-0286-6_2, © The Author(s) 2014

item-based CF algorithm, because it employs items' similarities for the formation of the neighborhood of nearest users. A pitfall of CF is the cold start problem: new items have received only few ratings, so they cannot be recommended; new users have performed only few transactions, so other users similar to them can be hardly found.

Content-based filtering assumes that each user operates independently. It exploits only information derived from documents or item features (eg. terms or attributes) [4, 20, 23]. In particular, it exploits a set of attributes, which describes the items and recommends other items similar to those that exist in the user's profile. This way, the cold start problem for new items and new users are alleviated, *provided* that users prefer items that are similar in content to those they have already chosen. However, the pitfall of CB is that there is no diversity in the recommendations. That is, the user gets recommendations that are very familiar to her, since the recommended items are similar to those in her item profile.

Hybrid algorithms attempt to combine CB with CF. The combination of content with rating data helps capture more effective correlations between users or items, which yields more accurate recommendations. The Fab System [1], combines CB and CF in its recommendations, by measuring similarity between users after first computing a profile for each user. Fab initially categorizes documents by a CB filter and then recommends them to the test user based on his relevance feedback. In contrast, the CinemaScreen System [25] runs CB on the results of CF. In particular, the CinemaScreen system computes predicted rating values for movies based on CF and then applies CB to generate the recommendation list.

Finally, apart from blending the content with rating data, Social web has allowed the emergence of new data combinations that can provide even more robust recommendations [22]. For instance, social networks such as Facebook, LinkedIn, etc., include information about the connections (link data) between humans. There are two main types of recommendations in social networks. The first one is related to the link prediction task, whereas the second one refers to the rating prediction and item recommendation task.

2.2 Definitions and Basic Factors

In this section, we identify the major factors that critically affect all CF algorithms, since they are among the most popular methods in recommender systems. Our analysis focuses on the basic operations of the CF process, which consists of three stages.

- *Stage 1*: formation of the user or item neighborhood with objects of similar ratings and behavior.
- *Stage 2*: generation of a top-N list with algorithms that construct a list of best item recommendations for a user.
- *Stage 3*: quality assessment of the top-N list.

Table 2.1 Factors affect CF algorithms

Factor name	Short description	Stage
Sparsity	Limited percentage of rated products	1
Scalability	Computation increase by the number of users and items	1
Train/test data size	Data are divided into training and evaluation or test	1,3
Neighborhood size	Number of neighbors used for the neighborhood formation	1
Similarity measure	Measures that calculate proximity of two objects	1
Recommendation list size	Number of top-N recommended items	2
Recommendation list creation	Algorithms for the top-N list generation	2
Positive rating threshold	Positive and negative ratings segregation	2,3
Evaluation Metrics	Metrics that evaluate the quality of top-N list	3
Setting a Baseline method	A simple method against which performance is compared	3
Past/future items	The segregation between a priori known and unknown items	3

In the rest of this section we elaborate on the aforementioned factors, which are organized with respect to the stage where each one is involved. The examined factors,which are detailed in the following, are succinctly described in Table 2.1.

Table 2.2 summarizes the symbols used in the sequel. To ease the discussion, we will use the running example illustrated in Fig. 2.1, where U_{1-10} are users and I_{1-6} are items. As shown, the example data set is divided into a training and a test set. Null cells (no rating) are represented as zeros.

Note that, in addition, a few more factors have been identified, like the impact of subjectivity during the rating or issues related to the preprocessing of data [3,19,27]. Nevertheless, we do not examine these factors, because their effect is less easily determinable.

2.2.1 First Stage Factors

Sparsity: In most real-world cases, users rate only a very small percentage of items. This causes data sets to become sparse. The problem of sparsity is extensively studied. In such cases, the recommendation engine cannot provide precise proposals, due to lack of sufficient information. A similar problem of CF algorithms is cold-start, as mentioned previously [21].

In a few works [19, 27], there is a preprocessing step that fills missing values. Several other works [11, 15, 28] focus only on very sparse data. Also, related work provides benchmark data sets with different sparsity, e.g., the Jester data

Table 2.2 Symbols and definitions

Symbol	Definition
k	Number of nearest neighbors
N	Size of recommendation list
$NN(u)$	Nearest neighbors of user u
$NN(i)$	Nearest neighbors of item i
P_τ	Threshold for positive ratings
$\mathcal{I}$	Domain of all items
$\mathcal{U}$	Domain of all users
u, v	Some users
i, j	Some items
$\mathcal{I}_u$	Set of items rated by user u
$\mathcal{U}_i$	Set of users rated item i
$r_{u,i}$	The rating of user u on item i
$\bar{r}_u$	Mean rating value for user u
$\bar{r}_i$	Mean rating value for item i
$p_{u,i}$	Predicted rate for user u on item i
$|T|$	Size of the test set
c	Number of singular values
A	Original matrix
U	Left singular vectors of A
S	Singular values of A
V'	Right singular vectors of A
A^*	Approximation matrix of A
$\mathbf{u}$	User vector
$\mathbf{u_{new}}$	Inserted user vector
n	Number of training users
m	Number of items

a

	I_1	I_2	I_3	I_4	I_5	I_6
U_1	4	1	1	4	0	0
U_2	1	4	4	0	0	4
U_3	2	1	4	0	0	4
U_4	1	2	1	1	0	0
U_5	4	1	2	0	1	0
U_6	1	5	4	0	4	0

b

	I_1	I_2	I_3	I_4	I_5	I_6
U_7	2	1	4	0	1	0
U_8	1	2	1	0	2	5
U_9	4	1	2	1	1	0
U_{10}	1	4	1	0	4	0

Fig. 2.1 (a) Training set ($n \times m$), (b) test set

set [9] is dense; in contrast the Movielens data sets [10] are relatively sparse. The degree of sparsity, however, depends on the application type. To provide particular conclusions, someone has to experiment with the amount of sparsity as well.

Scalability: Scalability is important, since the number of users/items is very large in real-world applications. Related work [27] has proposed the use of dimensionality reduction techniques, which introduce a trade-off between the accuracy and the execution time of CF algorithms.

Train/Test Data Size: There is a clear dependence between the training set's size and the accuracy of CF algorithms [28]. Additionally, after an upper threshold of the training set size, the increase in accuracy is less steep. However, the effect of overfitting is less significant compared to general classification problems. In contrast, small training set sizes impact accuracy in a negative way. Therefore, a fair evaluation of CF algorithms should be based on adequately large training sets. Though most related research uses a size around 80%, there exist works that use significantly smaller sizes [18]. From our experimental results we concluded that an 75% training set size constitutes an adequate choice. But we have to notice that the training/test size should not be data set independent. (In the running example, we set training size at 60%.)

Neighborhood Size: The number, k, of nearest neighbors used for the neighborhood formation results in a tradeoff: a very small k leads to low accuracy, because there are not enough neighbors to base the prediction. In contrast, a very large k impacts precision too, as the particularities of user's preferences can be blunted due to the large neighborhood size. In most related works [10, 26], k has been examined in the range of values between 10 and 100. The optimum k value depends on the data characteristics (e.g., sparsity). Therefore, for better tuning CF algorithms should be evaluated against varying k (in the running example, we have set $k = 3$).

Similarity Measure: Related works [11, 18, 19, 28] have mainly used Pearson correlation and cosine similarity. In particular, user-based (UB) CF algorithms use the Pearson correlation (Eq. (2.1)),[1] which measures the similarity between two users, u and v. Item-based (IB) CF algorithms use a variation of adjusted cosine-similarity (Eq. (2.2)),[2] which measures the similarity between two items, i and j, and has been proved to be more accurate [18, 28], as it normalizes the bias from subjective ratings.

$$\text{sim}(u, v) = \frac{\sum_{\forall i \in S} (r_{u,i} - \bar{r}_u)(r_{v,i} - \bar{r}_v)}{\sqrt{\sum_{\forall i \in S} (r_{u,i} - \bar{r}_u)^2} \sqrt{\sum_{\forall i \in S} (r_{v,i} - \bar{r}_v)^2}}, \quad S = I_u \cap I_v. \quad (2.1)$$

$$\text{sim}(i, j) = \frac{\sum_{\forall u \in T} (r_{u,i} - \bar{r}_u)(r_{u,j} - \bar{r}_u)}{\sqrt{\sum_{\forall u \in U_i} (r_{u,i} - \bar{r}_u)^2} \sqrt{\sum_{\forall u \in U_j} (r_{u,j} - \bar{r}_u)^2}}, \quad T = U_i \cap U_j. \quad (2.2)$$

---

[1] Means $\bar{r}_u, \bar{r}_v$ are the mean ratings of u and v over their co-rated items.

[2] Means $\bar{r}_u, \bar{r}_v$ are taken over all ratings of u and v.

a

	U_1	U_2	U_3	U_4	U_5	U_6
U_7	-0.19	0.19	1	-0.76	0.33	-0.14
U_8	-0.5	0.44	0.38	1	-0.82	0.67
U_9	0.41	-0.94	0.14	-0.47	1	-0.95
U_{10}	-0.5	0.5	-0.76	1	-0.82	0.67

b

	$1^{st}\,NN$	$2^{st}\,NN$	$3^{st}\,NN$
U_7	$U_3(1)$	$U_5(0.33)$	$U_2(0.19)$
U_8	$U_4(1)$	$U_6(0.67)$	$U_2(0.44)$
U_9	$U_5(1)$	$U_1(0.41)$	$U_3(0.14)$
U_{10}	$U_4(1)$	$U_6(0.67)$	$U_2(0.5)$

c

	I_1	I_2	I_3	I_4	I_5	I_6
I_1	-	-0.65	-0.66	0.36	-0.68	-0.42
I_2	-0.65	-	0.18	-0.51	0.50	-0.36
I_3	-0.66	0.18	-	-0.66	0.10	0.67
I_4	0.36	-0.51	-0.66	-	0	0
I_5	-0.68	0.50	0.10	0	-	0
I_6	-0.42	-0.36	0.67	0	0	-

d

	$1^{st}\,NN$	$2^{st}\,NN$	$3^{st}\,NN$
I_1	$I_4(0.36)$	-	-
I_2	$I_5(0.50)$	$I_3(0.18)$	-
I_3	$I_6(0.67)$	$I_2(0.18)$	$I_5(0.10)$
I_4	$I_1(0.36)$	-	-
I_5	$I_2(0.50)$	$I3(0.10)$	-
I_6	$I_3(0.67)$	-	-

Fig. 2.2 (**a**) Users' similarities matrix, (**b**) users' nearest neighbors in descending similarity matrix, (**c**) items' similarities matrix, (**d**) items' nearest neighbors in descending similarity matrix

Herlocker et al. [11] proposed a variation of the previous measures, which henceforth is denoted as Weighted Similarity (WS). If sim is a similarity measure (e.g., Pearson or cosine), then WS is equal to $\frac{\max(c,\gamma)}{\gamma} \cdot sim$, where c is the number of co-rated items.

Equation (2.1) takes into account only the set of items, S, that are *co-rated* by both users. This, however, ignores the items rated by only one of the two users. The number of the latter items denotes how much their preferences differ. Especially for the case of sparse data, by ignoring these items we discard significant information. Analogous reasoning applies for Eq. (2.2), which considers (in the numerator) only the set of users, T, that both co-rated the examined pair of items. The same applies for WS, which is based on Eqs. (2.1) or (2.2). To address the problem, in the following, we will examine alternative definitions for S and T.

The application of Pearson Correlation (Eq. (2.1)) to the running example is depicted in Fig. 2.2a and the resulting k-nearest neighbors(k-NN) are given in Fig. 2.2b. Respectively, the similarities between items, calculated with the adjusted cosine measure (Eq. (2.2)), are given in Fig. 2.2c, whereas Fig. 2.2d depicts the nearest neighbors. As only positive values of similarities are considered during the neighborhood formation, the items have different neighborhood size.

2.2.2 Second Stage Factors

Recommendation List's Size: The size, N, of the recommendation list results in a tradeoff: with increasing N, the absolute number of relevant items (i.e., recall) is expected to increase, but their ratio to the total size of the recommendation list (i.e., precision) is expected to decrease. (Recall and precision metrics are detailed in the following.) In related work [15, 28], N usually takes values between 10 and 50. (In the running example, we set $N = 2$.)

a

```
Vector GenTopMF(u, NN(u), I, N)
//User u
//Set NN(u)
//int I,N
begin
    for i = 1 to I
      f[i] = 0;
    foreach ν ∈ NN(u)
      for i = 1 to I
        if r[ν][i] ≥ Pτ
          f[i] = f[i] + 1;
    sort(f); //descending order
    n = 0; i = 1;
    while (n < N and f[i] > 0)
      A[n] = i;
      n = n + 1;
    return A;
end.
```

b

```
Vector GenTopMF(u, NN(i), I, N)
//User u
//Set NN(i)
//int I,N
begin
    for j = 1 to I
      f[j] = 0;
    for j = 1 to I
      if r[u][j] ≥ Pτ
        foreach ν ∈ NN(i)
          f[ν] = f[ν] + 1;
    sort(f); //descending order
    n = 0; i = 1;
    while (n < N and f[i] > 0)
      A[n] = i;
      n = n + 1;
    return A;
end.
```

Fig. 2.3 Generation of top-N list based on most frequent algorithm for (**a**) user-based algorithm and (**b**) item-based algorithm

Positive Rating Threshold: It is evident that recommendations should be "positive". It is not useful to recommend an item that will be rated with 1 in scale 1–5. Nevertheless, this issue is not clearly defined in several works. We argue that "negatively" rated items should not contribute to increasing the accuracy, and we use a rating-threshold, P_τ, to recommended items with rating no less than this value. If not a P_τ value is used, then the results may become misleading, since negative ratings can contribute to the measurement of accuracy.

Recommendation List Creation: The most often used technique for the generation of the top-N list counts the frequency of each item inside the found neighborhood, and recommends the N most frequent ones [26]. Henceforth, this technique is denoted as Most-Frequent item recommendation (MF). MF can be applied to both user-based and item-based CF algorithms. For example, assume that we follow the aforementioned approach for the test user U_7, for $k = 3$ and $N = 2$. For the case of user-based recommendation, the top-2 list includes items I_3, I_6. In contrast, for the case of item-based recommendation, the top-2 list includes 2 items of I_6 or I_2 or I_5 because all three have equal presence. Figure 2.3 describes the corresponding two algorithms (for the user and item-based CF, respectively).

We have to mention that these two algorithms, in contrast to the past work, in addition include the concept of positive rating threshold ($P\tau$). Thus, "negatively" rated items do not participate in the top-N list formation. Moreover, it is obvious that the generation of top-N list for the user-based approach is more complex and time consuming. The reason is that the former algorithm finds, firstly, user neighbors in the neighborhood matrix and then counts presences of items in the user-item matrix. In contrast, with the item-based approach the whole work is completed in the item neighborhood matrix.

Karypis [15] reports another technique, which additionally considers the degree of similarity between items. This takes into account that the similarities of the k neighbors may vary significantly. Thus, for each item in the neighborhood, this technique counts not just their number of appearances, but the similarity of neighbors as well. The N items with the highest sum of similarities are finally recommended. Henceforth, this technique is denoted as Highest-Sum-of-Similarities item recommendation (HSS). HSS is applicable only to item-based CF. For our example, the top-2 list based on this algorithm includes the items I_6, I_2, because they have the greater sum of similarities. Note that for both techniques, we have to remove from the recommendation list, these items that are already rated from the test user.

2.2.3 Third Stage Factors

Evaluation Metrics: Several metrics have been used for the evaluation of CF algorithms in related works [11, 12]: for instance the Mean Absolute Error (MAE) or the Receiving Operating Characteristic (ROC) curve. Although MAE has been used in most of these works, it has received criticism as well [18]. MAE is able to characterize the accuracy of prediction, but is not indicative of the accuracy of recommendation, as algorithms with worse MAE many times produce more accurate recommendations than others with better MAE. Since in real-world recommender systems the experience of users mainly depends on the accuracy of recommendation, MAE may not be the preferred measure. Other extensively used metrics are *precision* and *recall*. These metrics are simple, well known, and effective to measure the accuracy of recommendation procedure.

For a test user that receives a top-N recommendation list, let R denote the number of *relevant recommended items* (the items of the top-N list that are rated higher than P_τ by the test user). We define the following:

- *Precision* is the ratio of R to N.
- *Recall* is the ratio of R to the total number of relevant items for the test user (all items rated higher than P_τ by her).

Notice that with the previous definitions, when an item of the top-N list is not rated at all by the test user, it is considered as *irrelevant* and counts negatively to precision (as we divide by N). In the following, we also use $F_1 = 2 \cdot recall \cdot precision / (recall + precision)$. F_1 is used because it combines both the previous metrics.

Past/Future Data: In real-world applications, recommendations are derived only from the currently available ratings of the test user. However, in most of related works, all the ratings of each test user is considered a priori known. For a more realistic evaluation, recommendations should consider the division of the items of the test user into two sets [14]: (1) the past items of the test user, and (2) the future items of the test user.

2.3 Brief Literature Review

In 1992, the Tapestry system [8] introduced Collaborative Filtering (CF). In 1994, the GroupLens system [24] implemented a CF algorithm based on common users preferences, known as user-based CF algorithm, because it employs users' similarities for the formation of the neighborhood of nearest users. Since then, many improvements of user-based algorithm have been suggested, e.g., [3, 10].

In 2001, another CF algorithm was proposed. It was based on the items' similarities for a neighborhood generation of nearest items [15, 28] and is denoted as item-based CF algorithm.

Most related work followed the two aforementioned directions (i.e., user-based and item-based). Herlocker et al. [11] weight similarities by the number of common ratings between users/items. Deshpande and Karypis [6] apply item-based CF algorithm combined with conditional-based probability similarity and Cosine Similarity. Xue et al. [33] suggest a hybrid integration of aforementioned algorithms (memory-based) with model-based algorithms.

All aforementioned algorithms are memory-based. Their efficiency is affected from scalability of data. This means that they face performance problems, when the volume of data is extremely huge. To deal with this problem, many model-based algorithms have been developed [3]. However, there are two conflicting challenges. If an algorithm spends less execution time, this should not influence its quality. The best outcome would be to improve quality with the minimum calculation effort.

Furnas et al. [7] proposed Latent Semantic Indexing (LSI) in the area of Information Retrieval to deal with the aforementioned challenges. More specifically, LSI uses SVD to capture latent associations between the terms and the documents. SVD is a well-known factorization technique that factors a matrix into three matrices. Berry et al. [2] carried out a survey of the computational requirements for managing (e.g., folding-in[3]) LSI-encoded databases. They claimed that the reduced-dimensions model is less noisy than the original data.

Sarwar et al. [27, 29] applied dimensionality reduction for the user based CF approach. They also used SVD for generating predictions. In contrast to our work, Sarwar et al. [27, 29] do not consider two significant issues: (1) predictions should be based on the users' neighbors and not on the test (target) user, as the ratings of the latter are not a priori known. For this reason we rely only on the neighborhood of the test user. (2) The test users should not be included in the calculation of the model, because they are not known during the factorization phase. For this reason, we introduce the notion of pseudo-user to include a new user in the model (folding in), from which recommendations are derived. Other related work also includes Goldberg et al. [9], who applied Principal Components Analysis (PCA) to facilitate off-line dimensionality reduction for clustering the users, and therefore, achieves

---

[3]Folding in terms or documents is a simple technique that uses existing SVD to represent new information.

rapid on-line computation of recommendations. Hofmann [13] proposed a model-based algorithm, which relies on latent Semantic and statistical models. Moreover, Symeonidis et al. [30, 31] proposed a novel CF algorithm, which uses Latent Semantic Indexing (LSI) to detect rating trends and performs recommendations accordingly. Recently, Koren [16, 17], who is member of the winning team in the Netflix prize, proposed SVD++ method, which adds in the plain SVD also information taken from user/item bias and other implicit feedback. As the Netflix prize competition has demonstrated, matrix factorization models are superior to classic nearest-neighbor techniques.

2.4 Recommender Systems Paradigms

This section presents and briefly analyzes three Web Sites based on the three parallel approaches that have emerged in the context of recommender systems (i.e. CF, CB and hybrid methods).

CF exploits other users and their preferences to justify a recommendation to the target user. Usually the explanation is of the type "Customers who bought/rated item X also bought/rated items $Y, Z, \ldots$". It relies on the premise that both the target user and the users, which were used as an explanation have similar interests. A representative commercial recommender system that provides such recommendations is the online e-Commerce store Amazon.com.[4] As shown in Fig. 2.4, the user is presented with similar items that other customers have chosen to buy. The system assumes that the user is viewing an item, which they are already interested in. The Amazon system then finds similar users, who have already bought that specific item, and recommends the items that they bought to the user.

Fig. 2.4 Recommendation in Amazon

---

[4]http://www.amazon.com

Our Justified Recommendations			
[Movie id]	[Movie title]	[The reason is]	[because you rated]
1526	Witness (1985)	Ford, Harrison (I)	21 movies with this feature
1273	Color of Night (1994)	Willis, Bruce	7 movies with this feature
1004	Geronimo: An American Legend (1993)	Hackman, Gene	7 movies with this feature
1442	Scarlet Letter, The (1995)	Oldman, Gary	7 movies with this feature
1044	Paper, The (1994)	Close, Glenn	7 movies with this feature
693	Casino (1995)	De Niro, Robert	6 movies with this feature
274	Sabrina (1995)	Pollack, Sydney	6 movies with this feature
1092	Dear God (1996)	Kinnear, Greg	5 movies with this feature

Fig. 2.5 Recommendations in MoviExplain

One example of the combination of content with rating data is MovieExplain[5] [32], which combines the rating user profile and the feature item profile to reveal the favorite features of users. MoviExplain builds a feature profile for each user and provides as explanation, the feature that influenced most a recommendation, showing also how strong is this feature in the feature profile of a user. As shown in Fig. 2.5, the link "The reason is" reveals the favorite feature that influenced most the MoviExplain's recommendations, whereas the link "because you rated" shows how strong is this feature in the feature profile of a user.

In recent years, new innovations in online Social Networks have encouraged more sharing between users even of different networks. The recommendations are based on the common network that two users belong to. The most striking of these innovations is Facebook Login (formerly Facebook Connect). The way it works is that partner firms install Login buttons and plugins on their websites and devices, which give Facebook users automatic access to information about their friends' activities. Such an example is the HuffPost Social News. HuffPost[6] is a site run by the Huffington Post, a well known American blog, where Facebook users can see what their friends have been reading and exchange stories and comments about them. The system's user interface is shown in Fig. 2.6.

The personalized HuffPost Social News pages create a forum for users to converse about news stories they have read, and in some cases add their relevant information for Facebook friends to read. The Huffington Post creates a social news experience with the "Recommendations" plugin on its home page, showing users personalized recommendations along with explanations. The system's user interface is shown in Fig. 2.6. The post recommendations (i.e. blog stories) are explained based on both the preferences of a user's friends and the ratings that these items have received.

---

[5]http://delab.csd.auth.gr/MoviExplain
[6]http://www.huffingtonpost.com/

2.5 Social and Economic Report

The basic premise of recommenders is to reduce noise and filter out information, which is not relevant to the user taste. From the social point of view, the use of recommenders help people gain access to products or services that match their tastes. However, recommenders enforce users to insert information in log files of third party servers, arising privacy issues.

Regarding economy, there are several paradigms such as the Amazon.com, the Netflix.com and the Google.com recommendation engines, which have been proved as highly profitable. In particular, Amazon.com [5] claims that 35 % of products sales result from recommendations. Moreover, almost 66 % of movies rented in Netflix.com are recommended and Google News Recommendations generate 38 % more click-throughs [5].

In the same direction, friend recommendations in social networks stimulates users to expand their social circle, which is absolutely critical to retaining them and increasing the power of the network itself. Eventually, this brings income to a social network, based on the fact that companies invest their marketing budget except from target markets as well in mass power social networks. Recently, the incorporation of location to the recommendation process allows businesses to investigate new ways of profit. For instance, Foursquare.com incorporated a venue recommender in its mobile app. This recommender aims at offering to businesses a great advertising channel using their location, distance and users check-in history logs to stimulate users to visit the place.

References

1. M. Balabanovic, Y. Shoham, Fab: content-based, collaborative recommendation. Commun. ACM **40**(3), 66–72 (1997)
2. M. Berry, S. Dumais, G. O'Brien, Using linear algebra for intelligent information retrieval. SIAM Rev. **37**(4), 573–595 (1994)
3. J. Breese, D. Heckerman, C. Kadie, Empirical analysis of predictive algorithms for collaborative filtering, in *Proceedings of the 14th Conference on Uncertainty in Artificial Intelligence (UAI)*, Madison, WI (1998), pp. 43–52
4. R. Burke, Hybrid recommender systems: survey and experiments. User Model. User-adapt. Interact. **12**(4), 331–370 (2002)
5. O. Celma, P. Lamere, Music recommendation tutorial, in *International Conference on Music Information Retrieval (ISMIR 2007)*, Vienna (2007)
6. M. Deshpande, G. Karypis, Item-based top-n recommendation algorithms. ACM Trans. Inf. Syst. **22**(1), 143–177 (2004)
7. G. Furnas, S. Deerwester, S. Dumais, Information retrieval using a singular value decomposition model of latent semantic structure, in *Proceedings of the 13th ACM SIGIR International Conference on Research and Development in Information Retrieval (SIGIR)*, Grenoble (1988), pp. 465–480
8. D. Goldberg, D. Nichols, M. Brian, D. Terry, Using collaborative filtering to weave an information tapestry. Commun. ACM **35**(12), 61–70 (1992)
9. K. Goldberg, T. Roeder, T. Gupta, C. Perkins, Eigentaste: a constant time collaborative filtering algorithm. Inf. Retr. **4**(2), 133–151 (2001)
10. J. Herlocker, J. Konstan, A. Borchers, J. Riedl, An algorithmic framework for performing collaborative filtering, in *Proceedings of the 22th ACM SIGIR International Conference on Research and Development in Information Retrieval (SIGIR)*, Berkeley, CA (1999), pp. 230–237
11. J. Herlocker, J. Konstan, J. Riedl, An empirical analysis of design choices in neighborhood-based collaborative filtering algorithms. Inf. Retr. **5**(4), 287–310 (2002)
12. J. Herlocker, J. Konstan, L. Terveen, J. Riedl, Evaluating collaborative filtering recommender systems. ACM Trans. Inf. Syst. **22**(1), 5–53 (2004)
13. T. Hofmann, Latent semantic models for collaborative filtering. ACM Trans. Inf. Syst. **22**(1), 89–115 (2004)
14. Z. Huang, H. Chen, D. Zeng, Applying associative retrieval techniques to alleviate the sparsity problem in collaborative filtering. ACM Trans. Inf. Syst. **22**(1), 116–142 (2004)
15. G. Karypis, Evaluation of item-based top-n recommendation algorithms, in *Proceedings of the 10th International Conference on Information and Knowledge Management (CIKM)* (2001), pp. 247–254
16. Y. Koren, Collaborative filtering with temporal dynamics, in *Proceedings of the 15th ACM SIGKDD International Conference on Knowledge Discovery and Data Mining (KDD)*, Paris (2009), pp. 447–456
17. Y. Koren, Collaborative filtering with temporal dynamics. Commun. ACM **53**(4), 89–97 (2010)
18. R. McLauglin, J. Herlocher, A collaborative filtering algorithm and evaluation metric that accurately model the user experience, in *Proceedings of the 27th ACM SIGIR International Conference on Research and Development in Information Retrieval (SIGIR)*, Sheffield (2004), pp. 329–336
19. B. Mobasher, H. Dai, T. Luo, M. Nakagawa, Improving the effectiveness of collaborative filtering on anonymous web usage data, in *Proceedings of the IJCAI Workshop on Intelligent Techniques for Web Personalization (ITWP)*, Seattle, WA (2001), pp. 53–60
20. R. Mooney, L. Roy, Content-based book recommending using learning for text categorization, in *Proceedings of the 5th ACM Conference on Digital Libraries (DL)*, San Antonio, TX (2000), pp. 195–204

21. M. O'Mahony, N. Hurley, N. Kushmerick, G. Silvestre, Collaborative recommendation: a robustness analysis. ACM Trans. Internet Technol. **4**(4), 344–377 (2004)
22. A. Papadimitriou, P. Symeonidis, Y. Manolopoulos, A generalized taxonomy of explanation styles for traditional and social recommender systems. Data Min. Knowl. Discov. **24**(3), 555–583 (2012)
23. M. Pazzani, D. Billsus, Adaptive web site agents. Auton. Agent Multi Agent Syst. **5**(2), 205–218 (2002)
24. P. Resnick, N. Iacovou, M. Suchak, P. Bergstrom, J. Riedl, Grouplens: an open architecture for collaborative filtering on netnews, in *Proceedings of the ACM Conference Computer Supported Collaborative Work (CSCW)*, Chapel Hill, NC (1994), pp. 175–186
25. J. Salter, N. Antonopoulos, Cinemascreen recommender agent: combining collaborative and content-based filtering. Intell. Syst. Mag. **21**(1), 35–41 (2006)
26. B. Sarwar, G. Karypis, J. Konstan, J. Riedl, Analysis of recommendation algorithms for e-commerce, in *Proceedings of the ACM Conference on Electronic Commerce (EC)*, Minneapolis, MN (2000), pp. 158–167
27. B. Sarwar, G. Karypis, J. Konstan, J. Riedl, Application of dimensionality reduction in recommender system - a case study, in *Proceedings of the ACM SIGKDD Workshop on Web Mining for E-Commerce - Challenges and Opportunities (WEBKDD)*, Boston, MA (2000)
28. B. Sarwar, G. Karypis, J. Konstan, J. Riedl, Item-based collaborative filtering recommendation algorithms, in *Proceedings of the 10th International Conference on World Wide Web (WWW)*, Atlanta, GA (2001), pp. 285–295
29. B. Sarwar, G. Karypis, J. Konstan, J. Riedl, Incremental singular value decomposition algorithms for highly scalable recommender systems, in *Proceedings 5th International Conference on Computer and Information Technology (ICCIT)*, Dhaka (2002), pp. 27–28
30. P. Symeonidis, A. Nanopoulos, A. Papadopoulos, Y. Manolopoulos, Scalable collaborative filtering based on latent semantic indexing, in *Proceedings of the 21st AAAI Workshop on Intelligent Techniques for Web Personalization (ITWP)*, Boston, MA (2006), pp. 1–9
31. P. Symeonidis, A. Nanopoulos, A. Papadopoulos, Y. Manolopoulos, Collaborative recommender systems: combining effectiveness and efficiency. Expert Syst. Appl. **34**(4), 2995–3013 (2008)
32. P. Symeonidis, A. Nanopoulos, Y. Manolopoulos, Moviexplain: a recommender system with explanations, in *Proceedings of the 3rd ACM Conference on Recommender Systems (RecSys)*, New York, NY (2009), pp. 317–320
33. G. Xue, C. Lin, Q. Yang, W.S. Xi, H.J. Zeng, Y. Yu, Z. Chen, Scalable collaborative filtering using cluster-based smoothing, in *Proceedings of the 28th ACM SIGIR International Conference on Research and Development in Information Retrieval (SIGIR)*, Salvador (2005), pp. 114–121

Chapter 3
Online Social Networks

This chapter provides: (1) some definitions and basic concepts for Online Social Networks (OSNs), (2) a brief literature review of OSNs, (3) some paradigms of commercial OSNs, and (4) the transition of OSNs towards location-based services as an auxiliary dimension. Finally, the social and economic report of commercial OSNs helps the reader to realize the huge potential that Location-based Social Networks (LBSNs) have, based on the fact that OSNs have incorporated the location dimension in recent years.

3.1 Definitions

In this section, we annotate well-known fundamental definitions for Social Networks. These definitions derive from different scientific disciplines, showing the multidisciplinary scientific nature of OSNs. Moreover, to introduce OSNs smoothly, we further mention some useful concepts.

In the real physical world, a Social Network is a structure that consists of entities (i.e. people, companies etc.) and the ties between them. In other words, the aggregation of a person's social relationships yields his social network. In recent years, this structure has been delivered to the Internet, creating the core for OSNs. Specifically, existing social networking services in the web, act and in some cases substitute a physical social network structure. To achieve that, they allow users to build their own profiles containing personal info such as age, hobbies or political and religious views. After the user has created a profile, she is able to connect with other entities that have also a profile in the same social networking service. More than 50 years ago, researchers from different scientific fields were trying to understand the dynamics of the ties between people in real world. In 1967, Milgram [13] coined the term "small-world" after the results of his experiment have pointed out that the American society is characterized by short-length path connections. In particular, based on his provocative "small world" experiments, he claimed that everyone in the

P. Symeonidis et al., *Recommender Systems for Location-based Social Networks*, 21
SpringerBriefs in Electrical and Computer Engineering,
DOI 10.1007/978-1-4939-0286-6_3, © The Author(s) 2014

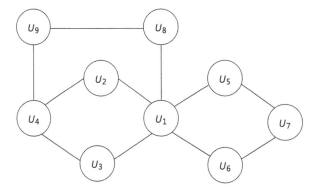

Fig. 3.1 Social network example

world could be connected to anyone else via "six degrees of separation" [13]. That is, for a randomly chosen pair of individuals, there exists with high probability a short chain of intermediaries that connect them, where "short" is usually interpreted to the logarithm of the population size.

As far as the OSNs are concerned, Garton et al. [5] defined an OSN as follows: "When a computer network connects people or organizations, it is a social network. Just as a computer network is a set of machines connected by a set of cables, a social network is a set of people connected by a set of social relationships, such as friendship, co-working or information exchange". Moreover, Boyd and Ellison [2] define OSNs as follows: "They are web-based services that allow individuals to: (1) construct a public or semi-public profile within a bounded system, (2) articulate a list of other users with whom they share a connection, and (3) view and traverse their list of connections and those made by others within the system. The nature and nomenclature of these connections may vary from site to site". Finally, Watts and Strogatz [20] arrived to a conclusion that: "In many real-world networks the probability of a tie between two actors is much greater if the two actors in question have another mutual acquaintance, or several."

A personal social graph is a graphic representation of all social links that a person has. Overall, a social graph is a drawing that plots the structure of interpersonal relations in a group situation and depicts all personal relations. This has been referred as: *"The global mapping of everybody and how they are related"*.

Figure 3.1 represents a single social graph containing nine nodes and the ties between them. Every node represents a person. Virtually, it can be observed that user U_1 is directly connected to users U_2, U_3, U_5, U_6, U_8.

Based on the assumption confirmed by [20], that people tend to connect with others similar to them, non-directly connected nodes are likely to build a connection. Thus, OSN providers are up and coming to perform actions, such as recommendations, based on the analysis of a user's social-graph plot.

For example, Facebook.com or Hi5.com use the following style of recommendation for recommending new friends to a target user U_1: "People you may know:

Fig. 3.2 Facebook's People You May Know friend recommendation

(1) user U_7 because you have two common friends (user U_5 and user U_6), (2) user U_9 because you have one common friend (user U_8) ...". The list of recommended friends is ranked based on the number of common friends each candidate friend has with the target user. Moreover, based on this recommendation style, a target user gets along with the friend recommendations the reasoning behind it. That is, a target user can see the reason (i.e. number of common friends) that influenced the recommendation process. As shown in Fig. 3.2, Facebook offers a "People You May Know" list of users which are possible friends of a target user, based on his existing connections with other users. The reasoning behind the recommendations is also shown, by providing the mutual friends that the two users share. For instance, in Fig. 3.2, the first possible friend of the target user is She-Rae because they share four friends, Joshua, Dennis, Frog and Bill. The target user can then add any person she may know from the proposed list to her list of friends.

3.2 Brief Literature Review

In this section, we review related work on link (i.e. Friend recommendation) and rating prediction (i.e. item/location/activity recommendation) in OSNs.

3.2.1   Link Prediction

The research area of *link prediction* in unipartite social networks, tries to infer which new interactions among members of a social network are likely to occur in the near future. There are two main approaches [10] that handle the link prediction problem. The first approach is based on local features of a network, focusing mainly on the nodes structure. There is a variety of local similarity measures such as Adamic and Adar index [1], Friend of a Friend (FOAF) algorithm [3], Preferential attachment [10] etc. The second approach is based on global features, detecting the overall path structure in a network. There is a variety of global approaches, such as Random Walk with Restart algorithm [14], the Katz status index [9] etc. Finally, Symeonidis et al. [15] proposed the FriendTNS algorithm to provide more accurate friend recommendations. They defined a transitive node similarity measure in OSNs by taking into account local and global features of a social graph.

Besides the aforementioned link prediction algorithms that are based solely on single-type graph structure, there are also other methods, which exploit also other data sources such as messages among users, co-authored papers, common tagging etc. For instance, a novel user interface widget has been proposed for providing users with recommendations of people [7]. These people recommendations were based on aggregated information collected from various sources across IBM organization (i.e. common tagging, common link structure, common co-authored papers etc.). An evaluation of four recommender algorithms (Content Matching, Content-plus-Link, FOAF algorithm and, SONAR) to help users discover new friends on IBM's OSN has been proposed in [3]. Recently, Lu et al. [11] proposed supervised link prediction using multiple heterogeneous sources (i.e. auxiliary networks) of information. However, the latter method focuses only on path counts and does not exploit other features and network characteristics, which can be informative for link formation (i.e. local graph characteristics).

3.2.2   Rating Prediction

In this section, we review related work on item recommendation in social networks. There are several methods that combine information from unipartite and bipartite graphs, focusing in the rating prediction (i.e. item/group recommendation) problem [6, 8, 19]. For example, TidalTrust [6] and MoleTrust [12] combine the rating data of collaborative filtering systems with the link data of trust-based social networks to improve the item recommendation accuracy. In particular, TidalTrust [6] performs a modified breadth first search in the trust network to compute a rating prediction. Furthermore, MoleTrust [12] considers paths of friends to a user-defined maximum-depth. Recently, Vasuki et al. [19] proposed affiliation/group recommendations based on the friendship network among users, and the affiliation/group network between users and groups. In particular, they suggested two models of user-community affinity for the purpose of making affiliation recommendations: one based on graph proximity, and another using latent factors to model users and communities. A probabilistic matrix factorization technique with trust propagation for leveraging item recommendations in social networks has been proposed in [8]. In particular, a model-based approach for item recommendation in social networks has been explored, employing matrix factorization techniques, by incorporating the mechanism of trust propagation into their model. Finally, Symeonidis et al. [17, 18] introduced a generalized framework that exploits multi-modal social networks to provide item and friend recommendations in social rating networks.

3.3 OSNs Commercial Paradigms

This section presents and briefly analyzes two commercial social networks. Except from visualizing real market, this section aims to depict the future potential of Location-based Social Networks, based on merits of their ancestors.

3.3.1 Facebook

Facebook, launched in February 2004, is the biggest social network nowadays. Although Facebook is not the core object in this book, it is presented as a commercial social networking service, so that the reader has a clear understanding of the most successful OSN and its features. It is obvious that the combination of these features along with other ideas brought Facebook to the top. The following description introduces some of the successful Facebook features:

Profile creation: Facebook gives users the opportunity to create a rich profile with a variety of information. Users can declare their sex, birthday, languages spoken, music and sports preferences, other interests, education, political and religious views. The previous attributes reflect user's tastes and preferences.

Connecting: This option enables user to search, find and send friend requests to other users to watch their updates in news feeds. In Facebook, two people can mutually agree to be listed as friends, to share information items such as photos, news, etc. Friendship links are initiated by one of the two people. For example, person *A* might find person *B* in the OSN and request to add him as a friend. Person *B* then receives both an email and a notification on the actual OSN interface informing him of the pending friend request. She can then either accept the undirected friendship, or choose to reject it. In case she rejects the friend request, the user who initiated the request is never informed of her rejection but instead her friend request is simply shown as pending.

Update status: This feature enables a user to generate content to be posted in the news feeds panel of her friends.

Chatting: The instant messaging option enables Facebook users to interact with friends, directly, through a chat box.

Like: Enables a user to express her countenance for friends or firms status updates. Like, is also used by people to follow firms through liking their fun page. The latter feature enables a user to get feeds provided by the organizations in the news feeds panel.

Event Creation: Facebook users can create event pages and invite their friends. Moreover they can have a forecast of their events attendance, monitoring who will attend the event, who is about to attend and who is not attending at all. Facebook gives guests the option, either to attend, maybe attend or decline the event invitation.

Groups: Every Facebook user is able to create a group and invite people in that. Groups could be open or closed for other users. For example, anyone can create a group and invite others to join. A group can represent a certain school or university, a sports team, a research group, etc. Two persons may belong in the same group or appear in pictures, which are tagged with their names. The same applies to videos that may include various users.

Facebook Apps and Login: Facebook enables other applications to be integrated in it. Facebook Login (formerly Facebook Connect) allows partner firms to install Login buttons and plug ins on their web sites and devices, which give Facebook users automatic access to information about their friends' activities. For instance, a user could easily integrate his Twitter into Facebook. The previous integration enables Facebook and Twitter to communicate in a way that user generated content in Twitter could be automatically posted in Facebook.

Check-in: This is a procedure related to the core subject of our book. However, check-in enables user to search nearby places and declare his presence there. This check-in action can be visible to user's friends.

Tagging: Social Tagging is the process by which many users add metadata in the form of keywords, to annotate and categorize items (songs/pictures, web links, products etc.). This is a feature that enables a user to tag others, including people and organizations, in her personal generated content. For instance, when a user uploads a picture, she can tag her friends in the picture and they will be notified about the upload of the picture.

Notice that, except the aforementioned basic features, Facebook provides users with auxiliary features to boost interaction between them, such as poke, birthday calendar, friend or fun page recommendations etc. Figure 3.3 presents the graphical user interface of Facebook, which consists of four basic columns. The left column shows the user groups, apps, messages and events. The middle column shows the news feeds provided by friends or liked fun pages. The right column shows the birthday calendar, recommendations, upcoming events, app requests and sponsored ads for users.

Fig. 3.3 Facebook's GUI

3.3.2 Twitter

Twitter, created in March 2006, is another type of social networking service. It is actually classified into the micro blogging category. Twitter services enable users to "follow" other users as well as sending short messages, known as "tweets". Following a user means, that every time she "tweets", in other words posts a public message, then any follower user will be notified. Users are allowed to group posts by topic or type, reply to posts, or repost messages from another user. In addition, twitter users are able to use it via the website using a computer, or a tablet or other device that connects to the web. In some countries there is a possibility to "tweet" via a mobile phone using an SMS service. Compared to Facebook, Twitter's social graph is directed and there is reciprocal friendship. Moreover, it is less noisy in terms of the provided features. It provides users fewer features:

Profile creation: This is an obligatory step for a user to access the provided services.

Tweet action: Tweet action is similar to the user's update status of Facebook. This function enables user to update her status in a micro-blogging way, utilizing a maximum of 140 characters. Moreover she can add photos, links or locations in her tweets.

Follow: This option enables a user to search, find and follow other users or organizations to watch their updates in news feeds.

Fig. 3.4 Twitter's GUI

Retweet action: This action enables a user to duplicate tweets from other users.
The duplicated tweet shows up in user's tweets, keeping the original generator as
the one who tweeted and the user, as the one who re-tweeted i.e. duplicated the
tweet.

Reply: Reply is an action used to reply a tweet by another user. The replied tweet
is shown up in both users news feeds so as other followers can see the source
user and the one who re-tweeted.

#Hashtag: The # symbol, called a hashtag, is used to mark keywords or topics in
a Tweet. It has been created organically by Twitter users as a way to categorize
messages.

Figure 3.4 depicts the Twitter GUI inside a user profile. Unlike Facebook,
the Twitter GUI comprises of only two columns. In the left column, Twitter
recommends people to follow and shows worldwide trends in terms of hashtags
or simple words. In the right column, a user can monitor the tweets generated by the
people she follows.

3.4 Social and Economic Report

User preferences change over time. Technological progressions, environmental changes etc. lead people to change their beliefs and habits to adopt the evolution. As a logical sequence, young people act different compared to their forefathers. A strong argument on the latter, is the distribution of internet use between different age groups. Table 3.1 presents the generation segmentation according to Evans et al. [4] in terms of ages.

Generation Y grew up in a time of immense and fast-paced change including virtually full employment opportunities for women, dual-income households as the standard, wide array of family types seen as normal, and computers in the home and schools. They were born in a technological and electronic society with global boundaries becoming more transparent. Generation Z is the newest generation and for the time being, these individuals are in their early formative years. They face global terrorism, the aftermath of 9/11, school violence, economic uncertainty, recession, and the crisis in US and EU. Generations Y and Z, are categorized as internet-based generations. Especially, Generation Z has never lived without the Internet [4]. They are both accustomed to high-tech and multiple information sources, with messages bombarding them from all sides. They prefer to socialize themselves through OSNs, which is a more secure environment and have become a great, digital, place for them to perform their actions. It is worth mentioning that Generations Y and Z will actually shape the development of such services. Conversely, OSNs will follow the development of those Gens to conserve the most multitudinous target group that have been accumulated in recent years.

We should also notice that there exist significant paradigms of the OSNs' power to change even a social structure, which can be considered as revolutionary for the humanity. For example, OSNs have massively increased their number of users. This increase was even powerful to influence several social demonstrations around the globe, such as the Egyptian Revolution against Mubarak's government.[1] This social phenomenon is basically caused by the great power that OSN hold in providing communication in a unified global manner.

Regarding the economical aspect of OSNs, monetization can be achieved through B2B or B2C channels. The most representative model of B2B monetization remains

Table 3.1 Generations segmentation	Generation distinctive name	Born
	Matures	Before 1946
	Baby Boomers	1946–1964
	Generation X	1965–1977
	Generation Y	1977–1994
	Generation Z	After 1994

---

[1]http://www.nytimes.com/2012/02/19/books/review/how-an-egyptian-revolution-began-on-face book.html

Facebook ads.[2] On the other hand, LinkedIn premium accounts[3] work on a B2C background, raising money from users able to get premium features on the LinkedIn network.[4]

Facebook is one of the most popular online social networking web sites and it counts more than 1 billion registered users, more than 500 millions of active users, 50 % of them log on to Facebook in any given day and it is translated into 70 languages.[5] As far as Twitter is concerned, it is estimated to have more than 200 million users. According to Radicati research group [16], social network users will reach 2.3 billion by 2016. In a more detailed approach, Wikipedia offers a list of all the Social networking sites and the number of their registered users nowadays.[6] The previous references, are highly valuable in terms of realizing the great power that social networks possess in modern societies.

OSNs have solved the problem of integrated and unified communication. For example, imagine a person with 300 adjacent connections, whereas her connections provide a link to 30 other connections, etc. It's a fact that, this user can easily reach all of her 300 connections directly or indirectly through her PC monitor at a glance. Moreover, by using an OSN, users have online access in the multimedia content generated by their connections at any time. At the same time, they use additional services such as live chat, messages, video call to inform their peers with a simple update procedure that enables friends to see the message anytime and anywhere. In conclusion, OSNs introduced a new way of communication integrating and incorporating in their GUI several media such as text, images, videos, etc.

3.5 Recommender Systems for OSNs

OSNs contain gigabytes of data that can be mined to make predictions about who is a friend of whom. OSNs recommend other people to users based on their common friends. The reason is that there is a significant possibility that two users are friends, if they share a large number of common friends. The premise of these recommendations is that individuals might only be a few steps from a desirable social friend, but not realize it. Thus, friend recommendation services allow users to get to know one's friends of friends and hence expand their own social circle. However, besides the explicit friendship relations between the users, there are also

---

[2]http://www.facebook.com/advertising?campaign_id=252705056280&placement=broadMovieEx plainxtive=6608774532&keyword=facebookads&extra_1=16eee6e6-f13e-c509-aaf1-000014a2666c

[3]http://www.linkedin.com/company/linkedin/linkedin-premium-accounts-3430/product

[4]http://www.linkedin.com/

[5]Facebook's statistics included in this section were obtained from: http://www.facebook.com/press/info.php?statistics

[6]http://en.wikipedia.org/wiki/List_of_social_networking_websites

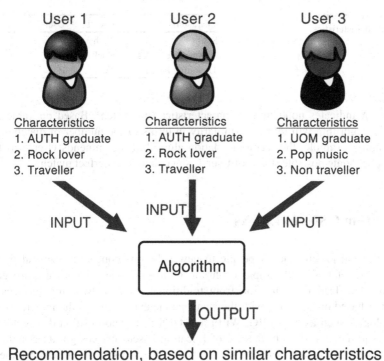

Fig. 3.5 Recommendation process

other implicit relations. For example, users can co-comment on products and they can co-rate products. Thus, item recommendation can be also provided is such systems based on the suggestions of our friends that we trust.

Recommender systems are widely used by OSNs to stimulate users to extend their personal social graph or for marketing purposes by recommending products that their friends have liked. Moreover, recommender systems can act like filters, trying to provide the right personalized information to each different user. Typically, a recommendation algorithm takes as input the preferences of the user and her friends from their profile and conscript them to yield recommendations for new ties such as friends, companies, places, products etc. Figure 3.5 shows the preferences of three different users in terms of education, music and traveling. Suppose that we want to recommend a new friend to User 1, whereas there are two possible suggestions (User 2 and User 3), which are both two hops far from the target user. Since no conclusion can be made from the social graph, the recommendation algorithm will try to compute similarities between the target user (i.e. User 1) and the other two users based on their preferences.

Table 3.2 represents deductively the algorithmic process to output a friend recommendation for User 1. The ✓ symbol depicts similarity match and thus qualifies a user to be recommended as a possible friend to the target User 1,

Table 3.2 Users with
different characteristics

Characteristic	User 1	User 2	User 3
Studies	Auth	✓	✗
Music preferences	Rock	✓	✗
Travelling habits	Yes	✓	✗

whereas ✗ indicates none similar characteristics (mismatch). Based on Table 3.2 data, the algorithm will infer the similarity in terms of the number of matches. In our running example, the recommendation algorithm recommends User 2 as the most appropriate possible friend of User 1, since there is a perfect match.

3.6   From OSNs to LBSNs

Technological progress leads people to new solutions concerning several fields. In the case of OSNs, the rapid evolution of technologies able to acquire geo-location data lead the transition from traditional social networking services to Location-based ones. Along with the smart phone evolution and the integration of technologies such as 3G or 4G, Wi-Fi and GPS in it, geo-position data got easy to be acquired. As a result, the use of Location-based services got accessible by almost every individual owning a smart phone. The fast growth of devices able to acquire geo-location data, led traditional OSN providers like Facebook and Twitter to add Location as a new dimension in the existing OSN structure. Social networks combined with geographical data, have evolved into location-based social networks (LBSNs). LBSNs such as Facebook Places, Google Places, Foursquare.com, etc., which allow users with mobile phones to contribute valuable information, have increased both in popularity and size. These systems are considered to be the next big thing on the web.

LBSNs allow users to use their GPS-enabled device, to "check in" at various locations and record their experience. In particular, users submit ratings or personal comments for the location/activity they visited/performed. That is, they "check in" at various places, to publish their location online, and see where their friends are. Moreover, they can either comment on a friend's location or comment on their own. These LBSN systems, based on a user's "check in" profile, can also provide activity and location recommendations. For an activity recommendation, if a user plans to visit some place, the LBSN system can recommend an activity (i.e. dance, eat, etc.). For a location recommendation, if a user wants to do something, the LBSN system can recommend a place to go. For example, Places is a Facebook feature that allows users to see where their friends are and share their location in the real world. Places uses the last place a user visited to determine which of her friends are nearby. Then, it sends to the user a number of location recommendations that are likely to be the most interesting to her.

Although traditional OSNs are successful, an LBSN proceeds a step ahead. Adding location dimension to OSNs is a more challenging task. Traditional graphical user interface of OSNs is considered old for LBSNs. The need of a map presence in the GUI of the LBSN is considered as a necessity, to visualize users' movements in the geo-locational dimension. Moreover, an OSN consists of one kind of nodes (unipartite graph), whereas an LBSN (except user-user graph) can be considered also as a bipartite (user-location) or even tripartite graph (user-location-activity). Furthermore, LBSNs are evolving in a very fast pace (through the daily user's check ins), demanding efficient database indexing and management. Finally, user's privacy in LBSN is even more important since user's location can be revealed.

References

1. L. Adamic, E. Adar, How to search a social network. Soc. Netw. **27**(3), 187–203 (2005)
2. D. Boyd, N. Ellison, Social network sites: definition, history, and scholarship. J. Comput. Mediat. Commun. **13**(1), 210–230 (2007)
3. J. Chen, W. Geyer, C. Dugan, M. Muller, I. Guy, Make new friends, but keep the old: recommending people on social networking sites, in *Proceedings of the 27th International Conference on Human Factors in Computing Systems (CHI)*, Boston, MA (2009), pp. 201–210
4. M. Evans, A. Jamal, G. Foxall, *Consumer Behaviour* (Wiley, London, 2006)
5. L. Garton, C. Haythornthwaite, B. Wellman, Studying online social networks. J. Comput. Mediat. Commun. **3**(1), 75–105 (1997)
6. J. Golbeck, Personalizing applications through integration of inferred trust values in semantic web-based social networks, in *Proceedings of the ISWC Semantic Network Analysis Workshop* (2005)
7. I. Guy, I. Ronen, E. Wilcox, Do you know?: recommending people to invite into your social network, in *Proceedings of the 13th International Conference on Intelligent User Interfaces (IUI)*, Sanibel Island, FL (2009), pp. 77–86
8. M. Jamali, M. Ester, A matrix factorization technique with trust propagation for recommendation in social networks, in *Proceedings of the 4th ACM Conference on Recommender systems(RecSys)*, Barcelona (2010), pp. 135–142
9. L. Katz, A new status index derived from sociometric analysis. Psychometrika **18**(1), 39–43 (1953)
10. D. Liben-Nowell, J. Kleinberg, The link prediction problem for social networks, in *Proceedings of the 12th International Conference on Information and Knowledge Management (CIKM)*, New Orleans, LO (2003), pp. 556–559
11. Z. Lu, B. Savas, W. Tang, I. Dhillon, Supervised link prediction using multiple sources, in *Proceedings of the 10th IEEE International Conference on Data Mining (ICDM)*, Sydney (2010), pp. 923–928
12. P. Massa, P. Avesani, Trust-aware collaborative filtering for recommender systems, in *Proceedings of Federated International Conference on the Move to Meaningful Internet (OTM): CoopIS, DOA, ODBASE*, Agia Napa (2004), pp. 492–508
13. S. Milgram, The small world problem. Psychol. Today **61**, 60–67 (1967)
14. J. Pan, H. Yang, C. Faloutsos, P. Duygulu, Automatic multimedia cross-modal correlation discovery, in *Proceedings of the 10th ACM SIGKDD International Conference on Knowledge Discovery and Data Mining (KDD)*, Seattle, WA (2004), pp. 653–658
15. A. Papadimitriou, P. Symeonidis, Y. Manolopoulos, Fast and accurate link prediction in social networking systems. J. Syst. Softw. **85**(9), 2119–2132 (2012)
16. Radicati Team, Social media market, 2012–2016. Technical Report, Radicati (2012)

17. P. Symeonidis, E. Tiakas, Y. Manolopoulos, Product recommendation and rating prediction based on multi-modal social networks, in *Proceedings of the 5th ACM Conference in Recommender Systems (RecSys)*, Chicago, IL (2011), pp. 61–68
18. P. Symeonidis, E. Tiakas, Y. Manolopoulos, A unified framework for link and rating prediction in multi-modal social networks. Int. J. Soc. Netw. Min. **1**(3/4) (2013)
19. V. Vasuki, N. Natarajan, Z. Lu, I.S. Dhillon, Affiliation recommendation using auxiliary networks, in *Proceedings of the 4th ACM Conference on Recommender Systems (RecSys)*, Barcelona (2010), pp. 103–110
20. D. Watts, S. Strogatz, Collective dynamics of small-world networks. Nature **393**(6684), 440–442 (1998)

Chapter 4
Location-Based Social Networks

Location-based Social Networks (LBSNs) can be considered as a special OSN category. Actually, an LBSN has the same OSN's properties, but qualifies location as the core object of its structure. This chapter initially provides some definitions and basic services that are offered by LBSNs. Next, a brief literature review, two commercial paradigms of LBSNs and a few location-based research projects are presented. Moreover, there is an economic and social report regarding LBSNs, which aims at investigating the field under a different, rather market-oriented prism. Finally, Sect. 4.5 provides an example of how a recommender system can be of benefit to an LBSN.

4.1 Definitions and Services of LBSNs

Recently, advances in broadband wireless networks and location sensing technologies led to the emergence of smart mobile phones, tablets etc. that allowed ubiquitous access to the Web. In this new era, users can benefit by getting ubiquitous access to location-based services from anywhere via mobile devices. Moreover, users can share location-related information with each other to leverage the collaborative social knowledge by using LBSNs.

Li and Chen [12] reported that LBSNs allow users to see where their friends are, to search location-tagged content within their social graph, and to meet others nearby. Zheng and Zhou [23] claimed that LBSNs consist of the new social structure made up of individuals connected by the interdependency derived from their locations in the physical world as well as their location-tagged media content, such as photos, video, and texts. According to Wikipedia,[1] LBSNs are a type of social networking in which geographic services and capabilities such as geo-coding

---

[1] http://en.wikipedia.org/wiki/Geosocial_networking

P. Symeonidis et al., *Recommender Systems for Location-based Social Networks*,
SpringerBriefs in Electrical and Computer Engineering,
DOI 10.1007/978-1-4939-0286-6_4, © The Author(s) 2014

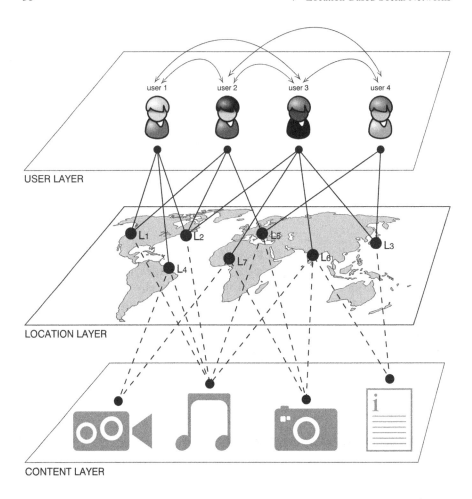

USER LAYER

LOCATION LAYER

CONTENT LAYER

Fig. 4.1 Visual representation of users, locations and content (i.e. photos/videos, tags, etc.)

and geo-tagging are used to enable additional social dynamics. As shown in Fig. 4.1, users can visit locations in the real world and can provide geo-tagged information content (e.g. comments, photos, videos). In particular, Fig. 4.1 presents three layers, namely, the user, the location, and the content layers. It is obvious that someone can exploit information from each layer independently to leverage recommendations. For instance, we can compute the geographical distance (i.e. Euclidean distance) between each pair of places in the location layer. Moreover, we can calculate the similarity among users based on the social network that exists in the user layer. Regarding the content layer, we can compute similarity among the information objects (i.e. video, tags etc.) based on their metadata. Please also notice the ternary relation among entities (i.e. user, location, content), which goes through all layers.

Acquiring this abundant contextual information, LBSNs can improve the quality of services on: (1) generic (non-personalized) recommendations of social events, locations, activities and friends, (2) personalized recommendations of social events, locations, activities and friends, and (3) user and group mobility behavior modeling and community discovery. In the remaining part of this section, we will briefly discuss the research work for each of the aforementioned categories of LBSNs services.

4.1.1 Generic Recommendations

Generic Recommendations compute the same recommendation list (location, activity, event etc.) for all users, regardless the personalized preferences of each individual user. The most simple recommender systems are those based on counting frequencies of occurrences or co-occurrences of some given dimension. For example, a simple recommender system could just count the number of check-ins per place, rank them and recommend those places with the larger number of check-ins.

A location recommender, for any user who travels in a specific city (e.g. New York), can first count each location's frequency of check-ins. Then, it can recommend the top-n locations by sorting these locations in decreasing order of their scores and selecting the n most popular. Notice that an interesting location can be defined as a cultural place, such as the Acropolis of Athens (i.e., popular tourist destinations), and commonly frequented public areas, such as shopping streets, restaurants, etc. As far as the activity recommendations is concerned, an activity recommender can provide a user with the most popular activities that may take place at a given location, e.g. dinning or shopping. Such a system that provides simple location and activity recommendations was the Gowalla.com web site, which was bought by Facebook in 2012. The system's user interface is shown in Fig. 4.2. As shown, a target user can provide to the system the activity she wants to do and the place she is (e.g. coffee in New York). Then, the system provides a map with coffee places, which are nearby the user's location (i.e. EuroPan Cafe in location A) and were visited many times (i.e. 17 times) from 16 people. All the aforementioned recommendations can guide a user in an unknown place of visit. Except the above baseline methods, there are other more advanced methods [16, 25] as well, which also provide generic (non-personalized) recommendations and will be discussed in Chap. 6.

4.1.2 Personalized Recommendations

The personalized recommender systems rely on past "check-in" history of users. Then, they correlate them with other users that have similar preferences and suggest to them new locations, activities and events. In particular, a personalized recommender exploits the time that someone has visited a location and her explicit ratings or comments on that location and predicts her interest in unvisited places. As

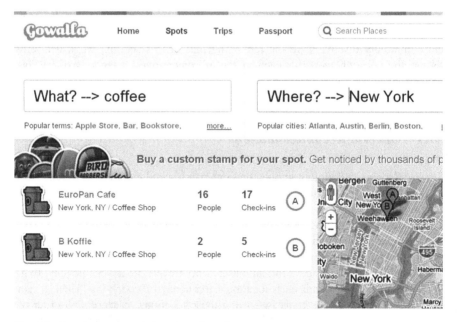

Fig. 4.2 Generic recommendations from Gowalla system

described in Sect. 2.1, there are three approaches that have emerged in the context of recommender systems: collaborative filtering (CF), content-based Filtering (CB) and hybrid methods. In the following, we briefly discuss, the special characteristics of each approach in the LBSN field.

CF methods recommend those locations, activities and events in a city to the target user, that have been rated highly by other users with similar preferences and tastes. In most CF approaches, only the locations and users' ratings are accessible and no additional information, i.e. locations or users, is provided. User-based CF, initially introduced in Sect. 2.1, employs users' similarities for the formation of the neighborhood of nearest users. User-based CF is an effective approach in terms of accurate recommendations. However, it cannot scale-up easily due to the high computation of user similarity matrix. In contrast, location-based CF algorithm employs locations' similarities for the formation of the neighborhood of nearest users, reducing the problem of scalability. In any case, a pitfall of both user-based and location-based CF is the cold start problem: new locations have received only few ratings, so they cannot be recommended; new users have performed only few visits, so there can be hardly found other users similar to them.

CB methods assume that each user operates independently. As a result, it exploits only information derived from location features. For example, a restaurant may have features such as cuisine and cost. If a user, in her profile, has set her preferable cuisine to be Chinese, then the Chinese restaurants will be presented to her. Apparently, the limitation of these systems lies upon the fact that other

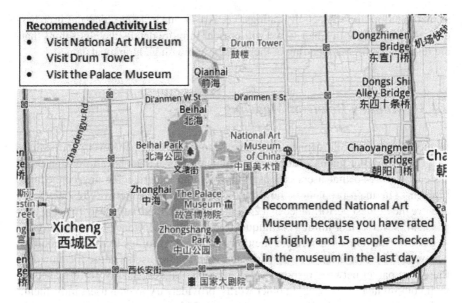

Fig. 4.3 Hybrid method of personalized recommendation

people's preferences are not considered. In particular, it exploits a set of attributes that describes the location and recommend other locations similar to those that exist in the user's profile. This way, the cold start problems, faced by CF methods, for new locations and new users are alleviated. However, the pitfall of CB is that there is no diversity in the location and activity recommendations.

The combination of social with geographical data, is becoming a way of handling shortcomings when only one type of data is taken into consideration. For example, the social graph (i.e. trust/friend connections) is not dealing with location analysis, whereas collaborative filtering maintains a user profile mainly based on rating data. The idea of a hybrid approach suggests that by using both data (i.e. social and rating data) it is possible to overcome each other's shortcomings and make the recommendation result to be more accurate. A hybrid system is the one mentioned in [25], where geographical data are combined with social data to provide location and activity recommendations. The authors in [25] use GPS location data, user ratings and user activities to propose recommendations to interested users along with appropriate explanations. The system's user interface is shown in Fig. 4.3. To use this system in activity recommendation, a user for example situated in Beijing, can input a location, as a location query; then the system can show the queried location on the map and suggest a ranking list of activities (top-3 here). These recommendations are then explained both by the previous visits of similar users at that location and the activity ratings, i.e. Art in our example, that the user has entered beforehand. In Chap. 6, we will present in more details the algorithms that are responsible for the blending of the social graph with geographical data.

4.1.3 Mobility Behavior Modeling

In recent years mobile connections have been established, for every individual, as an everyday life feature. Mobile communication providers have to design their systems stable, so that users can use them without experiencing any connection problem. The main problem is, how they will combine their network resources, to achieve the highest performance in a 24/7 manner, with minimum cost. To solve this problem, several data mining techniques process the user-generated mobile data, to discover movement patterns of the users and create profiles from their mobility behavior. Based on these users' profiles they can predict users' future location and thus, they parcel network resources to achieve the highest quality performance of their communication service. In other words, providers try to map users movement, including location, velocity and acceleration. This mapping procedure ensures providers that their network will be stable and properly working, or forewarn them for a possible, lack of resources and network overload, which may cause a network's block or drop.

To model users' mobility behavior, both analytical and simulation models can be used. Analytical mobility models are simplifying assumptions regarding the movement behavior of users by performing also simple mathematical calculations. In contrast, simulation models consider more detailed and realistic mobility scenarios. Such models can derive valuable solutions for more complex cases. There have been proposed [5] several simulation models for user's mobility modeling, such as the random way-point model, the random walk mobility model etc. Random way-point model [5] considers users moving randomly and freely without restrictions. Although, random models are simple and scalable, they have been shown to exhibit unrealistic users' behavior [10]. Recently, Dimokas et al. [8] proposed predictive techniques for location tracking. They considered the user's movement behavior as the outcome of an underlying stochastic process, which can be modeled by using well-established information-theoretic concepts and tools. Finally, authors of [1] based on the assumption that geography and social relationships are inextricably intertwined, carried out a research, using user-supplied address data and the network of associations between members of the Facebook social network. Engaging these measurements, they introduced an algorithm that predicts the location of an individual from a sparse set of located users with performance that exceeds IP-based geo-location. Notice that user and group mobility behavior modeling is beyond the scope of this book and we will no further discuss this research field.

4.2 Brief Literature Review

Since OSNs like Facebook and Google[+] have attracted millions of people, several algorithms have been designed to recommend friendship requests and advertisements based on the geographical position of users. Such systems (i.e. Gowalla.com, Foursquare.com, Facebook Places etc.) provide to users activity or location recommendations. There are also many applications that track people's mobility and get their "check-in" history to build a location-based recommendation [25].

As far as the user's location tracking, Backstrom et al. [1] use user-supplied address data and the network of associations between members of the Facebook social network to measure the relationship between geography and friendship. Using these measurements, they can predict the location of an individual. Scellato et al. [18] proposed a graph-based approach to study social networks with geographic information. They also applied new geo-social metrics to four large-scale Online Social Network data sets (i.e. Liveljournal, Twitter, FourSquare, BrightKite). In the same direction, Scellato et al. [19] presented a comprehensive study of the socio-spatial properties among users of LBSNs.

For location recommendations, Leung et al. [11] propose the Collaborative Location Recommendation (CLR) framework. This framework considers activities and different user classes to generate more precise and refined recommendations. The authors also incorporate a dynamic clustering algorithm, namely the Community-based Agglomerative-Divisive Clustering (CADC), into the framework to cluster the trajectory data into groups of similar users, similar activities and similar locations. The algorithm can also be updated incrementally when new GPS trajectory data are available.

There are also tensor-based approaches that have been proposed for geo-social recommendations because of the large dimensionality inherent of LBSNs (i.e. user, location, activity, etc.) For example, Biancalana et al. [3] implemented a social recommender system based on a tensor that is able to identify user preferences and information needs, and suggests personalized recommendations for possible Points of Interest. Furthermore, Zheng et al. [25] proposed a method, where geographical data are combined with social data to provide location and activity recommendations. The authors used GPS location data, user ratings and user activities to propose location and activity recommendations to interested users and explain them accordingly. Moreover, Zheng et al. [24] proposed a User Collaborative Location and Activity Filtering (UCLAF) system, which is based on tensor decomposition. In particular, they have added the user dimension and modeled their data with a three-dimensional tensor to provide targeted personalized recommendations using collaborative filtering techniques. Beside this, they used a model-based method that benefits from machine learning techniques, to predict missing values in aforementioned tensor structure. Similarly, the work in [26] presents a mobile recommendation system that, in addition to applying tensor factorization to the whole data set, incorporates two other algorithms to predict missing values. Finally, Symeonidis et al. [20] proposed a tensor-based approach that provides: (1) location and activity recommendations, and (2) friend recommendations by combining

FriendLink algorithm [13] with the geographical distance between users. Moreover, their tensor method includes an incremental stage, where newly created data is inserted into the tensor by incremental solutions [4].

4.3 Commercial and Research LBSN Paradigms

An example of a successful commercialization is Foursquare.[2] It was created in 2009; by September 2012, the company reported 25 million registered users, over 3 billion check-ins and an average of about 5 millions of check-ins per day. Users "check-in" at places using the foursquare mobile App. Each separate check-in awards them with points and sometimes with special "badges" (i.e. Newbie, Mayor, etc.). This is a part of a gamification process [7], which motivates users to check-in more and thus increase the use of the service. Moreover, they tend to augment their check-in behavior and try to get special offers, if they frequently visit a specific place. Thus, when a user checks-in in a place she is awarded with points. The number of points depends on several factors such as mayorships or "first of friends that checked in the place", etc. In addition, there is a relative ranking in Foursquare's application, depending on the points that friends of the user have accumulated. Figure 4.4 shows an example of a special offer which is given by Num Pang Sandwich Shop in New York to the Mayor of the place (the person which made the most check-ins lately) a special offer with "1 Grilled Com and 1 Blood Orange Lemonade".

Apart from commercialized LBSNs, many other academic research projects have emerged over the years. Figure 4.5 shows an approach named "I'm Feeling LoCo", proposed by Savage et al. [17]. The system learns user preferences by mining a person's social network profile. The physical constraints are delimited by the user's location and mean of transport, which are automatically detected through the use of a decision tree followed by a discrete hidden Markov chain. The novelty of this approach relies on the fusion of information derived from a user's social network profile and her mobile phone's sensors for geo-location discovery.

Takeuchi and Sugimoto [21] proposed CityVoyager, which introduces a novel real-world recommendation system, which makes recommendations of shops based on users' past location data history. As shown in Fig. 4.6, the system finds users' most visited shops and use them as input to the item-based collaborative filtering algorithm so as to make recommendations. In addition, they provide a method for further narrowing down the number of shops, based on prediction of user movements and geographical conditions of the city.

Finally, we notice that further information on other commercial and research mobile recommender systems, can be found in the paper by Ricci [14]. Another interesting survey on Recommendations in location-based social networks have been proposed by Bao et al. [2].

---

[2]http://foursquare.com/

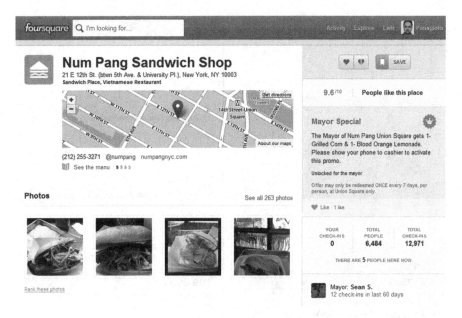

Fig. 4.4 Foursquare special offer to Mayor of a place

Fig. 4.5 LoCo interface

4.4 Social and Economic Report

LBSNs can increase customers' loyalty to companies, yielding better economic results for firms and changing the way people act in their everyday life. Using location as "input" to recommendation algorithms, LBSNs can achieve highly effective recommendations resulting to higher customers' engagement. It is worth mentioning how location bridges the gap between cyber and real world. Each time an individual checks-in a place, she reveals a preference or interest. It is clear that someone who checks-in every day in a basketball court is a basketball lover. Thus, an LBSN can recommend her other basketball lovers as friends or basketball courts as Points of Interest.

 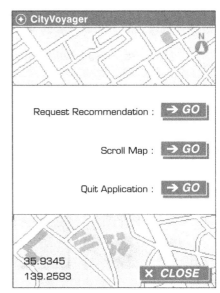

Fig. 4.6 Screenshots of CityVoyager

As far the social influence of LBSNs is concerned, Sakaki et al. [15] proposed an approach, able to inform users for an earthquake. The novelty of that approach stands on the speed that a new earthquake is being spread, using Twitter platform. In the same direction, De Longueville et al. [6] exploited information from an LBSN to acquire spatio-temporal data on forest fires. Users of LBSNs certainly gain more and more in terms of services, enforcing the opinion that LBSNs will be the next big thing in the internet during the upcoming years. To support the latter, it is likely that new generations i.e. Y and Z (see Sect. 3.4) mainly use mobile devices to access social networks and media.

For the economic perspective, a big revenue stream for commercial LBSNs has not yet been occurred. Despite the huge number of users involved in LBSNs, a clear way to monetize this has not yet been found. Thus, providers focused in improving user's experience. For example, the increasing number of Foursquare users over the recent years, displayed by Fig. 4.7, clearly demonstrates the trend towards Location-based services. As shown, the number of registered users in Foursquare increased rapidly from December 2009 till March 2012. Nowadays, Foursquare is the biggest pure LBSN. Although it has accumulated more than 25 million users, it has not still a strong enough revenue stream.

Lately, Foursquare concluded to an agreement with American Express[3] to let users access exclusive lists, tips, offers and experiences concerning the things they are interested in. Apart from cooperations with other organizations, there also other

---

[3]https://www.americanexpress.com/

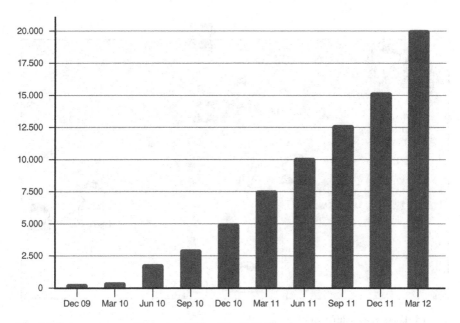

Fig. 4.7 Foursquare users December 2009 to March 2012

ideas for monetizing the location-based experience. For instance, Hsieh et al. [9] proposed a leisure guide that composes traveling paths from location traces of an LBSN. The commercialization of these paths is probably a feasible way to increase income for Location-based services. Additionally, the emerging work on Location-based recommendation systems increases the efficiency of personalized marketing [22] and as a result, the monetization sources.

4.5 Recommender Systems for LBSNs

Recommender systems, by using users' current location as an auxiliary source, can improve their recommendations about places or activities that a target user may be interested in. Generally, recommenders in LBSNs operate under the same framework as in traditional OSNs. The main difference is that in LBSNs the basic algorithmic input to generate a recommendation is user's location history. Specifically, the recommendation algorithm is trying to extract similarity using either the locations that users have been visited or the current users' location.

In the following, we will use a running example that points out deductively the process of recommendation in LBSNs. Figure 4.8, shows a global map with the locations that three users (Red, Black, Blue) have visited in the past. Let's assume that we want to recommend to the Red User new locations for traveling. Table 4.1 presents the users visits in places of our planet. As shown, both Red and Blue

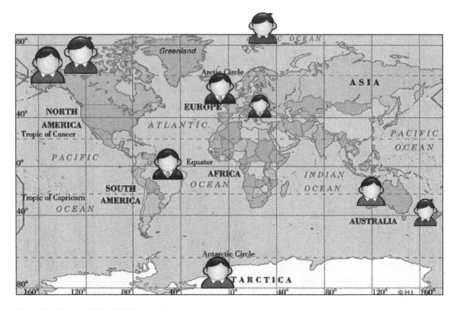

Fig. 4.8 Users visited different places

Table 4.1 Visits of users in places

Locations	Red User	Blue User	Black User
Alaska	Yes	Yes	–
Antarctica	Yes	–	–
Arctic	–	Yes	–
Australia	Yes	–	–
New Zealand	–	Yes	–
North Europe	–	–	Yes
South America	–	–	Yes
South Europe	–	–	Yes

users have visited Alaska. Thus, a recommendation algorithm considers them more similar in terms of traveling than the similarity between the Red and the Black users. If we had to recommend a place to the Red User, we would recommend either Antarctica or Oceania, which are places that the Blue User has visited in the past. Notice that there is also a semantic matching assigned to the locations that both Red user and Blue user checked-in. That is, the location semantics can infer two more matches i.e. Pole (Antarctic-Arctic), Oceania (Australia-New Zealand). It is important to emphasize that location semantics could also play an important role in a recommendation engine.

References

1. L. Backstrom, E. Sun, C. Marlow, Find me if you can: improving geographical prediction with social and spatial proximity, in *Proceedings of the 19th International Conference on World Wide Web (WWW)*, Raleigh, NC (2010), pp. 61–70
2. J. Bao, Y. Zheng, D. Wilkie, M.F. Mokbel, A survey on recommendations in location-based social networks. Technical Report in Microsoft Research Asia, http://research.microsoft.com/apps/pubs?id=191797, (2013)
3. C. Biancalana, F. Gasparetti, A. Micarelli, G. Sansonetti, Social tagging for personalized location-based services, in *Proceedings of the 2nd ACM CSCW International Workshop on Social Recommender Systems (SRS)*, Hangzhou (2011)
4. M. Brand, Incremental singular value decomposition of uncertain data with missing values, in *Proceedings of the 7th European Conference on Computer Vision (ECCV)*, Copenhagen (2002), pp. 707–720
5. T. Camp, J. Boleng, V. Davies, A survey of mobility models for ad hoc network research. Wirel. Commun. Mob. Comput. **2**(5), 483–502 (2002)
6. B. De Longueville, R. Smith, G. Luraschi, "omg, from here, i can see the flames!": a use case of mining location based social networks to acquire spatio-temporal data on forest fires, in *Proceedings of the International Workshop on Location Based Social Networks (LBSN)*, Seattle, WA (2009), pp. 73–80
7. S. Deterding, D. Dixon, R. Khaled, L. Nacke, From game design elements to gamefulness: defining "gamification", in *Proceedings of the 15th International Academic MindTrek Conference: Envisioning Future Media Environments (MindTrek)*, Tampere (2011), pp. 9–15
8. N. Dimokas, D. Katsaros, P. Bozanis, Y. Manolopoulos, Predictive location tracking in cellular and in ad hoc wireless networks. Mob. Intell. **69**, 163 (2010)
9. H.P. Hsieh, C.T. Li, Composing traveling paths from location-based services, in *Proceedings of the 6th International Conference on Weblogs and Social Media (ICWSM)*, Dublin (2012)
10. A. Jardosh, E.M. Belding-Royer, K.C. Almeroth, S. Suri, Towards realistic mobility models for mobile ad hoc networks, in *Proceedings of the 9th Annual International Conference on Mobile Computing and Networking (MobiCom)*, San Diego, CA (2003), pp. 217–229
11. K.W.T. Leung, D.L. Lee, W.C. Lee, Clr: a collaborative location recommendation framework based on co-clustering, in *Proceedings of the 34th ACM SIGIR International Conference on Research and Development in Information Retrieval (SIGIR)*, Beijing (2011), pp. 305–314
12. N. Li, G. Chen, Analysis of a location-based social network, in *Proceedings of the International Conference on Computational Science and Engineering (CSE'2009)*, vol. 4, Vancouver (2009), pp. 263–270
13. A. Papadimitriou, P. Symeonidis, Y. Manolopoulos, Friendlink: link prediction in social networks via bounded local path traversal, in *Proceedings of the 3rd Conference on Computational Aspects of Social Networks (CASON)*, Salamanca (2011), pp. 66–71
14. F. Ricci, Mobile recommender systems. Inf. Technol. Tourism **12**(3), 205–231 (2010)
15. T. Sakaki, M. Okazaki, Y. Matsuo, Earthquake shakes twitter users: real-time event detection by social sensors, in *Proceedings of the 19th International Conference on World Wide Web (WWW)*, Raleigh, NC (2010), pp. 851–860
16. M. Sattari, I. Toroslu, P. Senkul, M. Manguoglu, P. Symeonidis, Y. Manolopoulos, Geo-activity recommendations by using improved feature combination, in *Proceedings of the 4rd ACM SIGSPATIAL International Workshop on Location-Based Social Networks (LBSN)*, Pittsburgh, PA (2012), pp. 996–1003
17. N.S. Savage, M. Baranski, N.E. Chavez, I'm feeling loco: a location based context aware recommendation system, in *Proceedings of the 8th International Symposium on Location-Based Services (LBS)*, Vienna (2011), pp. 37–54
18. S. Scellato, C. Mascolo, M. Musolesi, V. Latora, Distance matters: geo-social metrics for online social networks, in *Proceedings of the 3rd Conference on Online Social Networks (WOSN)*, Boston, MA (2010), p. 8

19. S. Scellato, A. Noulas, R. Lambiotte, C. Mascolo, Socio-spatial properties of online location-based social networks, in *Proceedings of the 5th International Conference on Weblogs and Social Media (ICWSM)*, Barcelona (2011), pp. 329–336

20. P. Symeonidis, A. Papadimitriou, Y. Manolopoulos, P. Senkul, I. Toroslu, Geo-social recommendations based on incremental tensor reduction and local path traversal, in *Proceedings of the 3rd ACM SIGSPATIAL International Workshop on Location-Based Social Networks (LBSN)*, Chicago, IL (2011), pp. 89–96

21. Y. Takeuchi, M. Sugimoto, An outdoor recommendation system based on user location history, in *Proceedings of the 3rd International Conference on Ubiquitous Intelligence and Computing (UIC)*, Wuhan (2006), pp. 625–636

22. M. Wedel, R. Rust, T.S. Chung, Up close and personalized: a marketing view of recommendation systems, in *Proceedings of the 3rd ACM Conference on Recommender Systems (RecSys)*, New York, NY (2009), pp. 3–4

23. Y. Zheng, X. Zhou, *Computing with Spatial Trajectories* (Springer, Berlin, 2011)

24. V. Zheng, B. Cao, Y. Zheng, X. Xie, Q. Yang, Collaborative filtering meets mobile recommendation: a user-centered approach, in *Proceedings of the 24th AAAI Conference on Artificial Intelligence (AAAI)*, Atlanta, GA (2010), pp. 236–241

25. V. Zheng, Y. Zheng, X. Xie, Q. Yang, Collaborative location and activity recommendations with GPS history data, in *Proceedings of the 19th International Conference on World Wide Web (WWW)*, New York, NY (2010), pp. 1029–1038

26. V. Zheng, Y. Zheng, X. Xie, Q. Yang, Towards mobile intelligence: learning from GPS history data for collaborative recommendation. Artif. Intell. **184–185**, 17–37 (2012)

Part II
Recommendation Algorithms in LBSNs

Chapter 5
Framework

This chapter introduces the challenges that recommendation algorithms have to overcome in LBSNs. We also present the main algorithmic categories in the field of LBSNs (i.e. Collaborative Filtering, Semantically-enhanced, etc.). Moreover, we introduce the four types of recommendations in LBSNs (i.e. location, activity, friend, event). Finally, the reader meets an experimental framework for evaluating the quality of recommendations in LBSNs.

5.1 Algorithms' Challenges

As described in Sect. 3.6, recommendation algorithms for the mobile and social web (i.e. LBSNs) face more challenges compared with the recommendation algorithms for the traditional social web. Table 5.1 depicts some differences between traditional OSNs and LBSNs algorithms. In the following, we will briefly discuss each characteristic/requirement that recommendation algorithms should posses for use in LBSNs.

Need for More Compact GUI: The traditional graphical user interface of OSNs is considered as rather old for LBSNs. The need of a map presence in the GUI of the LBSNs is considered as a necessity, to visualize users' movements in the geo-locational dimension. Moreover, the phone display is limited (less room for display), which is a more challenging task for developers to visualize their service than traditional desktop GUI [7].

Node Heterogeneity: Graphs derived from LBSNs are more complicated. They can be unipartite (i.e. user-user), bipartite (i.e. user-location) or even tripartite graphs (i.e. user-location-activity). That is, there is a high percentage of node heterogeneity, which challenges recommendation algorithms [35].

Need for Scalability: LBSNs are evolving at a faster pace than traditional OSNs [35]. For example, it is common for users to visit new places without any

P. Symeonidis et al., *Recommender Systems for Location-based Social Networks*, 51
SpringerBriefs in Electrical and Computer Engineering,
DOI 10.1007/978-1-4939-0286-6__5, © The Author(s) 2014

Table 5.1 OSNs vs. LBSNs
algorithmic requirements

Algorithms' requirements	OSNs	LBSNs
Need for compact GUI	Medium	High
Node heterogeneity	Medium	High
User's privacy protection	Medium	High
Need for scalability	Medium	High
Need for fast response	Medium	High
Sparsity	Medium	High

prerequisites. In contrast, in an academic OSN the mobility of users to different-level conferences is not easy.

Need for Fast Response: LBSN users are usually on-the-go. They carry a resource-limited device (less storage and power supply). Hence recommendation results need to be chosen appropriately and delivered quickly [7]. The reason is that users have a shorter attention span since they are typically moving from place to place.

User's Privacy Protection: The nature of LBSNs imposes strong privacy barriers. The basic reason is that the imprinting of such sensitive information in the web, i.e. the user's physical location, raises questions for the safety of user's along several dimensions [2, 29]. For this reason, there is extensive related work regarding the user's privacy [12, 19, 21].

Sparsity: It refers to users not reluctant to provide their current location information. This causes data sets to become sparse. The problem of sparsity is broadly studied. In such cases, the recommendation engine cannot provide precise proposals, due to the lack of sufficient information. Another problem concerning the new incoming users or location is known as the "cold-start" problem. Again, the system has not adequate information to provide recommendation to new registered users or new inserted locations.

5.2   Categories of Recommendation Algorithms

As the size of data transferred from mobile devices augments rapidly, it is getting more complicated for researchers to deal with these massive data amounts. Dealing with this issue, recommender systems act like filters, by trying to provide the correct information to users and reduce the noise. For this purpose, recommendation services perform their action in three basic steps: First, they gather information. Then, they model it so that correlations between entities can be easily inferred and, finally, they provide the item recommendation lists. Nowadays, the increasing availability of mobile location acquisition technologies, such as GPS and/or WiFi, adds a new dimension in recommender systems; location is a new feature, which is acquired automatically through the location-aware devices. As a new dimension,

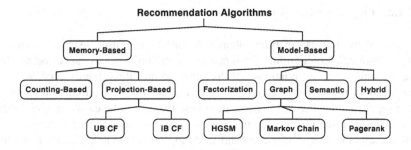

Fig. 5.1 Visual representation of the main algorithmic categories

location can easily bridge the gap between real and cyber/mobile world. Thus, implementations in sectors such as tourism or geo-located marketing can be absolutely useful and highly profitable [1, 31].

There are several algorithms that provide user, location or activities recommendations based on users mobile usage data. For example, there are systems that use collaborative filtering solutions, graph-based models, tensor-based methods, etc. All these algorithms can be categorized in two main categories:

- Memory-based algorithms, which perform computations directly on the entire database to identify the recommended items to a target user. These algorithms are cheap to compute and have proven to attain good results.
- Model-based algorithms, which recommend items to a target user after they have first computed a model. These algorithms have the overhead to build and update the model. Once they have build the model, they present good scalability.

Figure 5.1 shows a visual representation of the main algorithmic categories for the task of recommendation in LBSNs. The algorithmic families for recommendation in LBSNs have originated from different research fields and directions (data mining, intelligent systems, semantics etc.). In the next section, we refer in more details to the most representative research directions.

5.2.1 Memory-Based Algorithms

5.2.1.1 Counting-Based Techniques

The most simple methods are the ones based on counting frequencies of occurrences or co-occurrences of some given dimension (i.e. counting-based). For example, a simple recommender system could just count the number of check-ins per place, rank them and recommend those places with the larger number of check-ins. LBSNs that consider also ratings, can provide recommendations based on the average rating of each place or recommendations that take into consideration both frequency of visits and average rating. Such systems are typically regarded as baselines, since they are easily computed and operate with a minimal amount of data.

5.2.1.2   Projection-Based Techniques

Because of the ternary relations inherent in LBSNs (i.e. user, location, activity), many recommender algorithms designed to operate on matrices cannot be applied directly, unless ternary relations are decomposed into three binary relations (i.e. user-location, user-activity, activity-location). After the projection, Collaborative Filtering algorithms (i.e. user-based, item-based, etc.) can be easily applied in the user-location or the user-activity matrices. Notice that the activity-location projection discards the user information and leads to non-personalized location/activity recommendations.

Collaborative Filtering is a method to automatically predict (filter) the interests of a user by collecting preferences or taste information from many users (collaborating). A well honored research approach is user-based Collaborative Filtering [5,24], which forms neighborhoods based on similarities between users. In particular, for a test user, user-based CF employs users' similarities to form a neighborhood of his nearest users. Then, user-based CF recommends to the test user, the most frequent items in the formed neighborhood. Another algorithm proposed by Sarwar et al. [26], denoted as item-based Collaborative Filtering, forms item neighborhoods based on similarities between items. In the LBSNs field, a representative work utilizing this type of algorithm for location-based recommendation can be found in [39].

5.2.2   Model-Based Algorithms

5.2.2.1   Factorization Models

Since LBSNs are developed to host millions of users, it follows that there is crucial need for efficient data modeling and efficient storing. In that direction, researchers have proposed several model-based approaches [15, 17, 39] as parts of personalized and generic recommendation frameworks. This section advances the concept of data handling and aims at illustrating how data are represented and processed by algorithms to provide fast and accurate recommendation to users. In the following, we will present specific paradigms of how data are stored and modeled.

Data acquisition in LBSNs is achieved through GPS or Wi-Fi enabled devises. The accumulated data are stored in databases. However, LBSNs cache a steadily increasing users' check-in database. Thus, LBSNs should model the increasing incoming data efficiently to provide users with efficient recommendations. But how this massive amount of data is being organized to provide personalized recommendations? The review of the related work concludes to three different types of data modeling i.e. Matrix, Tensor and Graph based representation.

Matrix-based: The Singular Value Decomposition (SVD) is a method that factorizes a matrix or a tensor (i.e. a multi-dimensional matrix). Representative examples of matrix-based factorization approaches are the works of Zheng et al. [39] and

Fig. 5.2 Visual
representation of a
three-order tensor

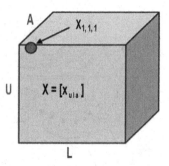

Sattari et al. [27]. They both proposed a data modeling approach that exploits several sources from users' GPS data. For example, Zheng et al. [39] exploits GPS trajectories to provide both location and activity recommendations. Their approach, denoted as Collaborative Location-Activity Filtering (CLAF), performs a factorization on the Location-Activity matrix. They also propagate information from two additional matrices (i.e. Location-Feature matrix and Activity-Activity similarity matrix) to reduce the data sparsity of the Location-Activity matrix.

Tensor-based: A tensor is a multidimensional matrix. Regarding LBSNs, they can be ideally modeled with a tensor structure, as they incorporate many participating entities/dimensions (i.e. users, locations, activities, tags, etc.). The interdependency of these dimensions is crucial to provide accurate recommendations to users. Thus, researchers have developed tensor-based approaches to tackle the challenge of exploiting also other contextual information such as user preferences, place rankings etc. Figure 5.2 shows a three-order tensor with dimensions (U),(L), and (A), which stand for Users, Locations and Activities, respectively. For instance, tensor's element X_{ula} means that user u has visited location l and performed activity a.

Multiverse Recommendation [8] is a method, which is based on the n-dimensional tensor factorization. This approach enables a generic and flexible integration of contextual information in a tensor, instead of performing a traditional two-dimensional matrix factorization. To achieve this, data are being modeled with an n-dimensional tensor having dimensions such as user, location and context (i.e. time, weather etc.). Zheng et al. [38] proposed also a method, known as PCLAF (personalized collaborative location and activity filtering), which is based on tensor decomposition. In particular, they used a three-dimensional tensor to represent relations between users, locations and activities. Based on this representation, they subsequently apply tensor decomposition.

In the same direction, Symeonidis et al. [17, 30] proposed a system that provides location, activity and friend recommendations by combining FriendLink algorithm [16] with the geographical distance among users. Their Incremental Tensor Reduction (ITR) model also includes an incremental stage, in which newly created data are inserted into the tensor by incremental solutions. Authors have

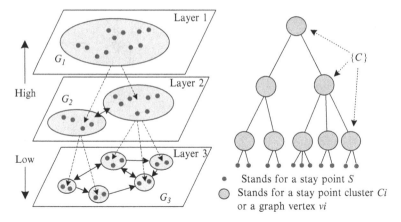

Fig. 5.3 A visual representation of a tree-based hierarchical graph adapted from [15]

built a Geo-Social recommender system prototype[1] that relies on user check-ins. The procedure involves selecting the location she is currently at, the activity she is performing there, and finally rating and commenting on that activity. Friends are recommended based on the FriendLink algorithm [16] and the average geographical distances among their "check-ins", which are used as link weights. Users, locations and activities are also inserted into a three-order tensor, which is then used to provide location and activity recommendations. This approach will be presented in more details in Chap. 8.

5.2.2.2 Graph-Based Models

Graphs are used to model data (i.e. user-user unipartite graph, user-location bipartite graph, user-location-activity tripartite graph etc.) and infer similarities among entities (i.e. users, locations, activities) in LBSNs. In the following, we introduce different graph-based models.

HGSM and TBHG Frameworks: Li et al. [15] proposed a framework to capture the similarity between users based on their location histories. This framework is known as Hierarchical Graph-based Similarity Measurement (HGSM) and is built on a tree-based hierarchical graph. Figure 5.3 depicts a tree-based hierarchical graph, where users' stay points are clustered into several spatial regions in a divisive manner. The similar stay points from various users are assigned to the same clusters on different layers. The model builds for each user a directed graph, in which a graph node is the cluster containing the user's stay points and a graph edge stands for the sequence of the clusters (geographic regions) being visited by this user. As shown in Fig. 5.3,

---

[1]http://delab.csd.auth.gr/geosocial2

a user's hierarchical graph (HG) can be formulated as a set of graphs $HG = \{G\}$ built on different geo-spatial scales. Each graph $G_i \in HG$ includes a set of vertices and edges, $G_i = (V, E)$, whereas $V = \{C\}$ is a set of clusters which contains the user's stay points.

HGSM models each individual's location history and effectively measures the similarity between users. This is due to the fact that it takes into account both the sequence property of people's movement behaviors and the hierarchy of the geographic spaces. Moreover, Zheng et al. [37] based on the tree-based hierarchical graph proposed TBHG, which is a HITS-based inference model [11] that considers an individual's access on a location as a directed link from the user to that location. In particular, TBHG model infers the interest of a location by taking into account the following three factors: (1) the interest for a location depends on users' travel experiences, (2) the users' travel experiences and location interests have a mutual reinforcement relationship, and (3) the interest for a location and the travel experience of a user are relative values and are region-related. Thus, TBHG takes into account a user's travel experience and the interest for a location in recommendation, so that only the locations that are really popular and also recommended by experienced users can be recommended.

Markov Chain: It is a mathematical structure that shows transitions from one state to another, between a finite or countable number of possible states. Quercia et al. [20, 22] proposed FriendSensing, which is a framework allowing mobile users to automatically discover their friends by using social network theories of "geographical proximity" and "link prediction" and by applying a Markov chain-based algorithm. Moreover, Geolife 2.0 [36] is an LBSN introduced by Microsoft Research Asia. It provides users with two types of location/activity recommendations (i.e. generic and personalized). Regarding the generic part, Geolife 2.0 uses a HITS-based algorithm to infer the popularity of the recommended locations/activities. Experimental results have shown that HITS-based [11] algorithm outperforms baseline methods, such as ranking location/activity recommendations by frequency. As far as the personalized recommendations are concerned, Geolife 2.0 uses a hierarchical graph-based similarity measure to model the individuals location history.

Bayesian Networks: It is a probabilistic graphical model that represents a set of random variables and their conditional dependencies via a directed acyclic graph (DAG). A representative work that uses Bayesian Network theory in LBSNs to provide recommendations in LBSN field is that of [18]. There also exist other methods that handle data as a Bayesian Network (BN). Park et al. [18] pre-processes geo-social information and trains the parameters of a BN. They obtain a conditional probability table (CPT) by performing the Expectation-Maximization (EM) model. For every new location/activity request by a user, the highest probability parameter, learned by the BN model, is selected. Finally, Ye et al. [34] developed a collaborative recommendation algorithm, which exploits geographical influence and incorporates it in a Naive Bayesian model.

Graph Clustering: Clustering is probably the most widely used method among the model-based approaches. The basic assumption behind clustering is that two users with similar preferences patterns are mechanistically related. Thus, it finds groups of users or items that have similar preferences. In particular, clustering is the task of assigning a set of objects into groups. The main concern of these algorithms is to group the more similar objects in the same cluster. Traditional clustering approaches, such as k-means and hierarchical clustering, put each user in exactly one cluster. A representative paradigm of such an algorithmic approach in LBSNs is Collaborative Location Recommendation framework (CLR) [14], which applies co-clustering on a user-activity-location tripartite graph, which is called Community Location Model or CLM graph. In particular, a co-clustering algorithm, namely the Community-based Agglomerative-Divisive Clustering (CADC) is applied on CLM to capture the relations between users, activities and locations. This model aims to mine similarities between the above mentioned entities and divides them into clusters of similar users, locations and activities. This approach will be presented in more details in Sect. 6.3.2.

5.2.2.3 Semantic-Based Models

Ye et al. [33] proposed a semantic annotation technique to automatically annotate all places with tags, which are crucial prerequisite for location search and recommendation services. The annotation algorithm learns a binary Support Vector Machine (SVM) classifier for each tag in the tag space to support multi-label classification. Based on the check-in behavior of users, they extract features of places from: (1) explicit patterns (EP) of individual places, and (2) implicit relatedness (IR) among similar places. The features extracted from EP are summarized from all check-ins at a specific place. The features from IR are derived by building a novel network of related places, where similar places are linked by virtual edges. Based on this network of related places, they determine the probability of a tag for each place by exploring the relatedness of places.

Xiao et al. [32] proposed a semantically-enhanced model, which computes similarities between users based on their GPS trajectories. That is, their approach first maps a user's GPS trajectory to a semantic location trajectory (SLH), e.g., shopping mall $\rightarrow$ restaurant $\rightarrow$ cinema. Then, they measure the similarity between different users' SLHs by using a maximal travel matching (MTM) algorithm. The advantage of their approach lies in two aspects. First, SLH carries more semantic meanings of a user's interests (beyond low-level geographic positions, i.e., geographic longitudes-amplitudes). Second, their approach can estimate the similarity between two users without overlaps in the geographic spaces, e.g., people living in different areas.

In the same direction, Cao et al. [6] proposed an approach that exploits location semantics for providing generic location recommendation in LBSN. This work introduces a general framework for mining significant semantic locations from

GPS data. In particular, they capture the latent relationships among locations in a unipartite graph and between locations and users in a bipartite graph. This approach will be presented in more details in Sect. 6.3.3.

5.3 Recommendation Types

Based on location data provided by users, LBSN providers can yield four different types of recommendations i.e. Friend, Location, Activity and Event recommendation. In the following, we introduce each one of these types.

5.3.1 Friend Recommendation

In LBSNs, two people can mutually agree to be listed as friends, to share information of places they visit, photos from these places, traveling routes in a city, etc. Regarding friend recommendation, as shown in Fig. 5.4, the target user is provided with a list of other users she may know. However, LBSNs usually do not exploit all different length paths of the network that connect a user with her potential friends. Instead, they consider only pathways of maximum length 2 between a user and his candidate friends. On the other hand, there are global approaches, which detect the overall path structure in a network, being computationally prohibitive for huge-size social networks. Symeonidis et al. proposed Friendlink algorithm [16], which provides recommendations by traversing paths (of bounded length, i.e. 3 or 4 hops) of a user's social graph. Thus, their friend recommendations, relies on the "algorithmic small world hypothesis".

In the same direction, Scellato et al. [28] studied the link prediction (i.e. friend recommendation) problem and concluded that about 30% of new links are added among "place-friends", i.e. among users who visit the same places. They have also showed that they can make the prediction space 15 times smaller, while still 66% of future connections can be discovered. Their approach is presented in more details in Sect. 6.4.2.

Fig. 5.4 Friend recommendation

Fig. 5.5 Location
recommendation

Fig. 5.6 Activity
recommendation

5.3.2 Location or Point of Interest (POI) Recommendation

When a user is in an unknown place, probably she wants to know where to head
for food or sightseeing. Location or POIs recommendation can provide helpful
guidance to a user for visiting interesting places, as shown in Fig. 5.5. Ye et al.
[34] argues that the geographical influence among POIs plays an important role
in user's check-in behavior and models it by means of a power law distribution.
Furthermore, he proposes a unified POI recommendation framework, which fuses
user's preference to a POI with social and geographical influence.

Another very recent study by Saez-Trumper et al. [25] has shown that recommending places that are closest to a user's geographic center of interest (e.g. user's
home), produces recommendations that are as accurate as item-based collaborative
filtering algorithm.

5.3.3 Activity Recommendation

An activity is usually consistent with the location that the user checks-in. For
example, when a user checks-in a basketball court she may perform activities such
as playing basketball, watching a basketball game or similar. Thus, a recommender
will suggest to the user activities related to basketball, as shown in Fig. 5.6.

Notice that there are approaches that exploit other contextual information (except
location) to provide activity recommendations for users. For instance, Bellotti et
al. [3] proposed the Magitti leisure guide. The proposed approach infers user
activity from context and patterns of user behavior. Using those information, without
any query generated by user, Magitti leisure guide automatically provides several
activity recommendations (i.e. eating, shopping, seeing, doing, or reading).

Fig. 5.7 Event
recommendation

5.3.4   Event Recommendation

This is a special case of activity recommendation, which recommends to a target
user several social events (concerts, parties, meetings, athletics events etc.) taking
place either in the digital world or in a specific city, as shown in Fig. 5.7. Compared
with the aforementioned types, event recommendation is relatively new. Many
recent research works aim at detecting events, such as concerts and festivals, using
geo-tagged information posted by a large number of users [13]. Intuitively, people
participating in such an event would post an increased amount of information at the
location where the event takes place. By mining the information that co-occurs in
particular locations, we can predict the existence of important geo-social events.

Quercia et al. [23] carried out a research about event recommendation in the
Greater Boston area. Specifically, they extracted social events that took place during
a certain period and combined them with obtained GPS coordinates from one
million mobile phone users. After analyzing the combined data set, they were able
to know which social event, residents of an area have attended. Upon this data, they
tried to provide generic event recommendations to users based only on their home
location (i.e., cold start users). They tested a variety of algorithms for recommending
social events (i.e. popular events, popular events in area, geographical close events,
etc.) Results of this study have shown that the most effective algorithm, is the
one that recommends events popular among residents of an area, whereas the least
effective algorithm recommends events that are geographical close to the area. For
personalized recommendations, Kayaalp et al. [9, 10] proposed a hybrid method,
which combines collaborative with content-based filtering to recommend events to
users of Facebook. This approach will be presented in more details in Sect. 6.4.3.

5.4   Evaluation

5.4.1   Metrics

In Sect. 2.2.3, we mentioned some basic metrics for evaluating the effectiveness
of recommender systems. We have noticed the Mean Absolute Error (MAE), the
receiver-operating characteristic ROC, and Precision-Recall as well, which originate
mainly from the Information Retrieval field. A more fair metric compared with
Precision-Recall is Mean Average Precision (MAP), which takes into consideration
the order of a hit in the recommendation list as well. That is, since we provide to a

test user u a top-k list of friends/location/activities/events, it is important to consider the order of the presented friends/locations/activities/events in this list. Let's assume that we recommend friends to a user. Then, it is better to have a correct guess in the first places of the recommendation list. Thus, we use the *Mean Average Precision (MAP)* to emphasize ranking of relevant users higher. We define MAP by Eq. (5.1):

$$MAP = \frac{1}{|N|} \sum_{u=1}^{|N|} \frac{1}{r_u} \sum_{k=1}^{r_u} Precision_u @k \qquad (5.1)$$

where N is the number of users in the test data set, r_u is the number of relevant users to a user u and $Precision_u @k$ is the precision value at the k-th position in the recommendation list for u. Notice that MAP takes into account both precision and recall and is geometrically referred as the area under the Precision-Recall curve. Notice that another metric similar to MAP is $nDCG$ (Normalized Discounted Cumulated Gain).

An equivalent metric to the receiver-operating characteristic (ROC) curve is the area under the ROC curve, known as AUC statistic that quantifies the accuracy of prediction algorithms and evaluates the improvement over pure chance. It is the probability that a correctly chosen existent friend/location has a higher similarity value than a randomly chosen non-existent friend/location. In the implementation, among n times of independent comparisons, if there are n' times the correct prediction friend/location has higher similarity value and n'' times the correct predicted friend/location and non existent friend/location have the same similarity value, then we define AUC by Eq. (5.2):

$$AUC = \frac{n' + 0.5 \times n''}{n} \qquad (5.2)$$

If all similarity values are generated from an independent and identical distribution, then the accuracy should be about 0.5. Therefore, the degree to which the accuracy exceeds 0.5 indicates the prevalence of the algorithm over pure randomness.

5.4.2 Protocols

5.4.2.1 Cross-validation Approach

Cross-validation is a alternative to the random sub-sampling. Each record is used the same number of times for training and exactly once for testing. The case study in Sect. 8 uses this kind of evaluation since it measures the performance of the recommenders within a real LBSN. In particular, it performs fourfold cross validation and the default size of the training set is 75%. For each user, it picks randomly 75% of his check-ins and friends. The task of all three recommendation types (i.e. friend, location, activity) is to predict the friends/locations/activities of the user's 25% remaining check-ins and friends, respectively.

5.4.2.2 Leave-One-Out Approach

This is a variant of the cross-validation approach. For location recommendations, one randomly picks, for each target user an activity performed by her. The task of the recommender is then to predict the locations the user has visited to perform this specific activity. This process is then repeated, each time with a randomly chosen activity per user, to further minimize the variance. Recall and precision values are then averaged over all the runs. The Leave-One-Out evaluation protocol changes analogously for activity recommendations.

5.4.2.3 Time-Based Split Approach

Another evaluation protocol for testing the recommendation quality is to split the data set by time. In this case, one chooses a point in time that lies within the time interval of the data set and then uses all check-in data before that point as training data and all check-in data after that point as test data. This technique resembles the situation of a recommender in a real LBSN, where only the previous check-ins are known at a specific recommendation time.

5.4.3 Datasets

There are not many data sets available for measuring the quality of recommendations in LBSNs. Quercia and Capra [20] used a data set from the Reality Mining Project at MIT. It contains collocation information from 96 users, which were collected via Bluetooth device discoveries. Furthermore, Zheng et al. [40] released a data set extracted from users' trajectories of the GeoLife prototype system. This data set, known as UCLAF data set, consists of 164 users, 168 locations and 5 different types of activities, including "Food and Drink", "Shopping", "Movies and Shows", "Sports and Exercise", and "Tourism and Amusement".

Papadimitriou et al. [17] extracted from their GeoSocialRec recommender web site,[2] a data set for the evaluation of their ITR algorithm. This data set initially consisted of 102 users, 46 locations and 18 different types of activities. However, as new data were inserted into the system, the data set is extended to

---

[2]http://delab.csd.auth.gr/geosocialrec

149 users, 436 locations and 112 activities. The extended data set[3] consists also of 594 friendship connections among users and 853 check-in quadruplets (i.e. user, location, activity, and rating information). There is also additional information about the geographical longitude and latitude of each location.

Recently some commercial LBSNs such as Foursquare, released API sets that allow scientists to crawl the check-in history of users and perform experiments. For example, Ye et al. [34] crawled the web sites of Foursquare and Whrrl by using their API sets, for a month to collect two data sets consisting of 153,577 users and 96,229 POIs in Foursquare, and 5,892 users and 53,432 locations in Whrrl, respectively. Berjani and Strufe [4] used crawled data from Gowalla, creating two data sets. The first comes from Austin in Texan (ATX) and the second from New York city (NYC). ATX consists of 11,896 users, 9,525 locations, 249,317 check-ins. NYC consists of 10,132 users, 9,290 places and 114,256 check-ins. In the same direction, Scellato et al. [28] downloaded four monthly snapshots of Gowalla data between May and August 2010 creating another data set to test their method. Finally, Leung et al. [14] used a data set that is collected from 50 different users' outdoor GPS trajectories, which were recorded from September 2009 to May 2010. This data set is used for the performance evaluation of their CLR recommendation framework.

References

1. ABI Research Market Analysis Report, 82 million location-based mobile social networking subscriptions by 2013. Technical Report, ABI Research, Nov 2008
2. Associated Press, Stalker victims should check for GPS. Technical Report, CBSNews.com, Feb 2003
3. V. Bellotti, B. Begole, E. Chi, N. Duchenaut, J. Fang, E. Isaacs, T. King, M. Newman, K. Partridge, B. Price, P. Rasmussen, M. Roberts, D. Schiano, A. Walendowski, Activity-based serendipitous recommendations with the Magitti mobile leisure guide, in *Proceedings of the 26th Annual SIGCHI Conference on Human Factors in Computing Systems (CHI)*, Florence (2008), pp. 1157–1166
4. B. Berjani, T. Strufe, A recommendation system for spots in location-based online social networks, in *Proceedings of the 4th Workshop on Social Network Systems (SNS)*, Salzburg (2011), pp. 4:1–4:6
5. J. Breese, D. Heckerman, C. Kadie, Empirical analysis of predictive algorithms for collaborative filtering, in *Proceedings of the 14th Conference on Uncertainty in Artificial Intelligence (UAI)*, Madison, WI (1998), pp. 43–52
6. X. Cao, G. Cong, C. Jensen, Mining significant semantic locations from GPS data. Proc. VLDB Endowment **3**(1–2), 1009–1020 (2010)
7. T. Horozov, N. Narasimhan, V. Vasudevan, Using location for personalized POI recommendations in mobile environments, in *Proceedings of the International Symposium on Applications on Internet (SAINT)*, Washington, DC (2006), pp. 124–129
8. A. Karatzoglou, X. Amatriain, L. Baltrunas, N. Oliver, Multiverse recommendation: n-dimensional tensor factorization for context-aware collaborative filtering, in *Proceedings of the 4th ACM Conference on Recommender Systems (RecSys)*, Barcelona (2010), pp. 79–86

---

[3]http://delab.csd.auth.gr/~symeon

9. M. Kayaalp, T. Ozyer, S.T. Ozyer, A collaborative and content based event recommendation system integrated with data collection scrapers and services at a social networking site, in *Proceedings of the International Conference on Advances in Social Network Analysis and Mining (ASONAM)*, Athens (2009), pp. 113–118

10. M. Kayaalp, T. Ozyer, S.T. Ozyer, A mash-up application utilizing hybridized filtering techniques for recommending events at a social networking site. Soc. Netw. Anal. Min. **1**(3), 231–239 (2011)

11. J. Kleinberg, Authoritative sources in a hyperlinked environment, in *Proceedings of the 9th Annual ACM-SIAM Symposium on Discrete Algorithms (SODA)*, San Francisco, CA (1998), pp. 668–677

12. B. Lee, J. Oh, H. Yu, J. Kim, Protecting location privacy using location semantics, in *Proceedings of the 17th ACM SIGKDD International Conference on Knowledge Discovery and Data Mining (KDD)*, San Diego, CA (2011), pp. 1289–1297

13. R. Lee, S. Wakamiya, K. Sumiya, Discovery of unusual regional social activities using geo-tagged microblogs. World Wide Web **14**(4), 321–349 (2011)

14. K.W.T. Leung, D.L. Lee, W.C. Lee, CLR: a collaborative location recommendation framework based on co-clustering, in *Proceedings of the 34th ACM SIGIR International Conference on Research and Development in Information Retrieval (SIGIR)*, Beijing (2011), pp. 305–314

15. Q. Li, Y. Zheng, X. Xie, Y. Chen, W. Liu, W.Y. Ma, Mining user similarity based on location history, in *Proceedings of the 16th ACM SIGSPATIAL International Conference on Advances in Geographic Information Systems (GIS)*, Irvine, CA (2008), pp. 34:1–34:10

16. A. Papadimitriou, P. Symeonidis, Y. Manolopoulos, Friendlink: link prediction in social networks via bounded local path traversal, in *Proceedings of the 3rd Conference on Computational Aspects of Social Networks (CASON)*, Salamanca (2011), pp. 66–71

17. A. Papadimitriou, P. Symeonidis, Y. Manolopoulos, Geo-social recommendations, in *Proceedings of the RecSys Workshop on Personalization on Mobile Applications (PeMA)*, Chicago, IL (2011)

18. M.H. Park, J.H. Hong, S.B. Cho, Location-based recommendation system using Bayesian user's preference model in mobile devices, in *Proceedings of the 4th International Conference in Ubiquitous Intelligence and Computing (UIC)*, Hong Kong (2007), pp. 1130–1139

19. K. Puttaswamy, N. Zhao, Preserving privacy in location-based mobile social applications, in *Proceedings of the 11th Workshop on Mobile Computing Systems and Applications (HotMobile)*, Annapolis, MD (2010), pp. 1–6

20. D. Quercia, L. Capra, Friendsensing: recommending friends using mobile phones, in *Proceedings of the 3rd ACM Conference on Recommender Systems (RecSys)*, New York, NY (2009), pp. 273–276

21. D. Quercia, S. Hailes, Sybil attacks against mobile users: friends and foes to the rescue, in *Proceedings of the 29th Conference on Information Communications (INFOCOM)*, San Diego, CA (2010), pp. 336–340

22. D. Quercia, J. Ellis, L. Capra, Using mobile phones to nurture social networks. IEEE Pervasive Comput. **9**(3), 12–20 (2010)

23. D. Quercia, N. Lathia, F. Calabrese, G. Di Lorenzo, J. Crowcroft, Recommending social events from mobile phone location data, in *Proceedings of the IEEE International Conference on Data Mining (ICDM)*, Sydney (2010), pp. 971–976

24. P. Resnick, N. Iacovou, M. Suchak, P. Bergstrom, J. Riedl, Grouplens: an open architecture for collaborative filtering on netnews, in *Proceedings of the ACM Conference Computer Supported Collaborative Work (CSCW)*, Chapel Hill, NC (1994), pp. 175–186

25. D. Saez-Trumper, D. Quercia, J. Crowcroft, Ads and the city: considering geographic distance goes a long way, in *Proceedings of the 6th ACM Conference on Recommender Systems (RecSys)*, Dublin (2012), pp. 187–194

26. B. Sarwar, G. Karypis, J. Konstan, J. Riedl, Item-based collaborative filtering recommendation algorithms, in *Proceedings of the 10th International Conference on World Wide Web (WWW)*, Atlanta, GA (2001), pp. 285–295

27. M. Sattari, M. Manguoglu, I.H. Toroslu, P. Symeonidis, P. Senkul, Y. Manolopoulos, Geo-activity recommendations by using improved feature combination, in *Proceedings of the ACM UbiComp International Workshop on Location-Based Social Networks (LBSN)*, Pittsburgh, PA (2012), pp. 996–1003

28. S. Scellato, A. Noulas, C. Mascolo, Exploiting place features in link prediction on location-based social networks, in *Proceedings of the 17th ACM SIGKDD International Conference on Knowledge Discovery and Data Mining (KDD)*, San Diego, CA (2011), pp. 1046–1054

29. B. Schilit, J. Hong, M. Gruteser, Wireless location privacy protection. IEEE Comput. **36**(12), 135–137 (2003)

30. P. Symeonidis, A. Papadimitriou, Y. Manolopoulos, P. Senkul, I. Toroslu, Geo-social recommendations based on incremental tensor reduction and local path traversal, in *Proceedings of the 3rd ACM SIGSPATIAL International Workshop on Location-Based Social Networks (LBSN)*, Chicago, IL (2011), pp. 89–96

31. The Economist: Editorial Team, A world of connections: a special report on networking. Technical Report, Economist (2010)

32. X. Xiao, Y. Zheng, Q. Luo, X. Xie, Finding similar users using category-based location history, in *Proceedings of the 18th ACM SIGSPATIAL International Conference on Advances in Geographic Information Systems (GIS)*, San Jose, CA (2010), pp. 442–445

33. M. Ye, D. Shou, W.C. Lee, P. Yin, K. Janowicz, On the semantic annotation of places in location-based social networks, in *Proceedings of the 17th ACM SIGKDD International Conference on Knowledge Discovery and Data Mining (KDD'2011)*, San Diego, CA (2011), pp. 520–528

34. M. Ye, P. Yin, W.C. Lee, D.L. Lee, Exploiting geographical influence for collaborative point-of-interest recommendation, in *Proceedings of the 34th ACM SIGIR International Conference on Research and Development in Information Retrieval (SIGIR)*, Beijing (2011), pp. 325–334

35. Y. Zheng, X. Zhou, *Computing with Spatial Trajectories* (Springer, Berlin, 2011)

36. Y. Zheng, Y. Chen, X. Xie, Y. Ma, GeoLife2.0: a location-based social networking service, in *Proceedings of the 10th International Conference on Mobile Data Management: Systems, Services and Middleware (MDM)*, Taipei (2009), pp. 357–358

37. Y. Zheng, L. Zhang, X. Xie, W.Y. Ma, Mining interesting locations and travel sequences from GPS trajectories, in *Proceedings of the 18th International Conference on World Wide Web (WWW)*, Madrid (2009), pp. 791–800

38. V. Zheng, B. Cao, Y. Zheng, X. Xie, Q. Yang, Collaborative filtering meets mobile recommendation: a user-centered approach, in *Proceedings of the 24th AAAI Conference on Artificial Intelligence (AAAI)*, Atlanta, GA (2010)

39. V. Zheng, Y. Zheng, X. Xie, Q. Yang, Collaborative location and activity recommendations with GPS history data, in *Proceedings of the 19th International Conference on World Wide Web (WWW)*, New York, NY (2010), pp. 1029–1038

40. V. Zheng, Y. Zheng, X. Xie, Q. Yang, Towards mobile intelligence: learning from GPS history data for collaborative recommendation. Artif. Intell. **184–185**, 17–37 (2012)

Chapter 6
Algorithms

This chapter provides more details on advanced research work proposed in LBSNs, and deepens in the algorithmic side of each method.

For generic recommendations, Zheng et al. [20] introduced Collaborative Location and Activity Filtering (CLAF). In the same direction, Sattari et al. [12] proposed Improved Feature Combination (IFC) algorithm. Notice that both CLAF and IFC use matrices for data representation.

For personalized recommendations, PCLAF [19] algorithm provides location and activity recommendations. Regarding the algorithmic framework, PCLAF is categorized in the Collaborative Filtering research field. PCLAF uses a tensor for data representation. Based on their previous works, i.e., CLAF and PCLAF, Zheng et al. [22] proposed RPCLAF [22], which uses Ranking within the recommendation process.

The research community has also proposed several other advanced algorithms for recommendation in LBSNs. For example, FriendSensing [8], RMF [1], CADC [6], incremental HOSVD [15], and USG [16] algorithms have introduced smart solutions for recommendation in LBSNs. In the following, there is a more detailed description of all the aforementioned and other selected approaches as well.

6.1 Matrix-Based Factorization

6.1.1 CLAF Algorithm

Zheng et al. [20] proposed CLAF, which provides a generic location recommendations within Geolife [21], a prototype LBSN proposed by Microsoft Research Asia.[1] In particular, they recommend activities to users for a given location and

---

[1] http://research.microsoft.com/en-us/labs/asia/default.aspx

P. Symeonidis et al., *Recommender Systems for Location-based Social Networks*,
SpringerBriefs in Electrical and Computer Engineering,
DOI 10.1007/978-1-4939-0286-6_6, © The Author(s) 2014

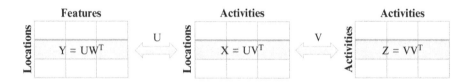

Fig. 6.1 Visual representation of Location-Feature, Location-Activity, and Activity-Activity matrices adapted from [20]

locations for an given activity. CLAF exploits in its model the relations between activities. That is, if a user goes to a cinema, then she may go to a restaurant, too. This is a kind of latent relationship that is aimed to be captured as activity correlation. Notice that CLAF incorporates information from both the features of the locations and the activity correlations. As shown in Fig. 6.1, CLAF uses additional information captured in Location-Feature and Activity-Activity matrices to predict missing entries in the Location-Activity matrix, which is usually very sparse.

CLAF is similar to Collective Matrix Factorization, originally proposed by Singh and Gordon [14]. The Collective Matrix Factorization model converts an objective function to an optimization problem, which is later solved iteratively by *Gradient Descent*.

6.1.2   IFC Algorithm

Sattari et al. [12] proposed an extended matrix model, namely Improved Feature Combination (IFC), that integrates data from one resource (e.g. Location-Activity matrix) with data from additional resources (e.g. Activity-Activity and Location-Feature matrices), to leverage generic Location/Activity recommendations. As shown in Fig. 6.2, the Location-Activity sub-matrix implies preference ratings of users on a specific activity in a specific location. Its entries actually correspond to the frequency of performing an activity in that location for all users. The Location-Activity sub-matrix contains available features of locations and the Activity-Activity sub-matrix represents relationships (semantic or others) among different activities.

IFC builds a model that injects additional information into the main data, by building an extended matrix. Then, on the merged data, Singular Value Decomposition (SVD) is applied to extract latent relations between locations, their features and activities that are performed by users. Several experiments have been conducted, and the results of IFC were compared with CLAF [20]. Their experiments showed that IFC outperforms CLAF in terms of prediction accuracy.

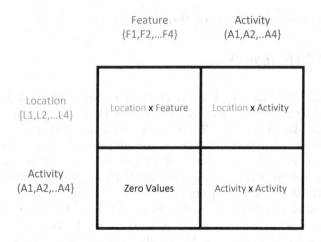

Fig. 6.2 Visual representation of an extended matrix, with Location-Feature, Location-Activity, and Activity-Activity sub-matrices

6.1.3 RMF Algorithm

Regularized matrix factorization [1], denoted as RMF, is another matrix-based model for personalized location recommendations in LBSNs. RMF applies collaborative filtering techniques on a dimensionally reduced User-Location matrix. Each matrix entry actually corresponds to the interest of a user in a location. RMF models users and locations as latent features. The predicted interest of a users u in a location l is calculated as:

$$\widehat{i_{ul}} = q_l^T \cdot p_u, \tag{6.1}$$

where q_l is a location-factors vector and p_u is a user-factors vector. In other words, the predicted interest $\widehat{i_{ul}}$ of a user u in a location l is inferred by the inner product of the aforementioned vectors, which should well approximate the real interest i_{ul} of a user. This approximation fitness is determined by minimizing the associated regularized squared error between the real and the predicted user's interest, which is solved iteratively through stochastic gradient descent. Notice that the factor vectors (i.e. location and user) are of much lower dimensionality compared to the original User-Location matrix and, hence, they can more easily be kept in main memory, speeding up the prediction time of RMF.

6.2 Tensor-Based Factorization

6.2.1 PCLAF Algorithm

Zheng et al. [19] introduced a personalized recommendation algorithm for
LBSNs, which performs Personalized Collaborative Location and Activity Filtering
(PCLAF). Unlike CLAF [20], PCLAF treats each user differently and uses a
collective tensor and matrix factorization to provide personalized recommendations.
As shown in Fig. 6.3, the novelty of PCLAF lays on the utilization of a
User-Location-Activity tensor along with the User-User, User-Location, Location-
Features and Activity-Activity matrices.

As also shown in Fig. 6.3, to fill missing entries in the tensor $\mathcal{A}$, PCLAF
decomposes $\mathcal{A}$ w.r.t. each tensor dimension (i.e. user, location, activity). Then,
PCLAF forces the latent factors to be shared with the additional matrices to utilize
their information. After such latent factors are obtained, PCLAF reconstructs an
approximation tensor $\hat{\mathcal{A}}$ by filling all the missing entries. Notice that PCLAF uses a
PARAFAC-style regularized tensor decomposition framework to integrate the tensor
with the additional matrices.

6.2.2 ITR Algorithm

Symeonidis et al. [15] proposed Incremental Tensor Reduction (ITR) to provide location/activity recommendations within an online prototype system, called

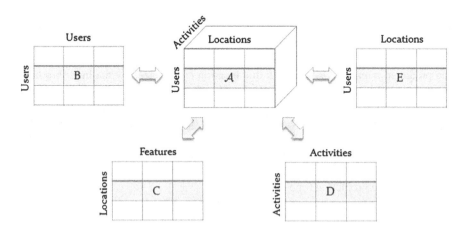

Fig. 6.3 Visual representation of a User-Location-Activity tensor along with the User-User, User-
Location, Location-Features and Activity-Activity matrices adapted from [19]

Geo-Social 2.0 recommender system.[2] ITR represents data by a three-order tensor (user, location, activity), on which dimensionality reduction is performed using the Higher Order Singular Value Decomposition (HOSVD) technique.

The tensor decomposition can be expressed as follows:

$$\hat{\mathcal{A}} := \hat{C} \times_u \hat{U} \times_l \hat{L} \times_a \hat{A} \tag{6.2}$$

where $\hat{U}, \hat{L}$, and $\hat{A}$ are low-rank feature matrices representing a mode, i.e., user, locations, and activities respectively, in terms of its small number of latent dimensions k_U, k_L, k_A, and $\hat{C} \in \mathbb{R}^{k_U \times k_L \times k_A}$ is the core tensor representing interactions between the latent factors. The model parameters to be optimized are represented by the quadruple $\hat{\theta} := (\hat{C}, \hat{U}, \hat{L}, \hat{A})$.

The basic idea of the HOSVD algorithm is to minimize an element-wise loss on the elements of $\hat{\mathcal{A}}$ by minimizing the square loss as shown by Eq. (6.3):

$$\underset{\hat{\theta}}{\text{argmin}} \sum_{(u,l,a) \in Y} (\hat{a}_{u,l,a} - a_{u,l,a})^2 \tag{6.3}$$

After the parameters are optimized, predictions can be done as:

$$\hat{s}(u,l,a) := \sum_{\tilde{u}} \sum_{\tilde{l}} \sum_{\tilde{a}} \hat{c}_{\tilde{u},\tilde{l},\tilde{a}} \cdot \hat{u}_{u,\tilde{u}} \cdot \hat{l}_{l,\tilde{l}} \cdot \hat{a}_{a,\tilde{a}} \tag{6.4}$$

where indices over the feature dimension of a feature matrix are marked with a tilde, and elements of a feature matrix are marked with a hat (e. g., $\hat{a}_{a,\tilde{a}}$).

In contrast to PCLAF, the ITR method uses incremental solutions to update the tensor, as more data are accumulated to the system. Compared to other solutions, ITR provides three types of recommendation, i.e., Location, Activity and Friend recommendation. The latter is achieved with the incorporation of Friendlink algorithm [7]. This approach will be presented in more details later in Chap. 8.

6.2.3 RPCLAF Algorithm

Zheng et al. [22] proposed Ranking-based Personalized Collaborative Location and Activity Filtering (RPCLAF). RPCLAF takes a direct way to solve the recommendation problem by using a ranking loss objective function. That is, instead of minimizing the prediction error between the real and the predicted user preference for an activity in a location (see Eq. (6.3)), the RPCLAF method formulates the user's location-activity pairwise preferences by Eq. (6.5):

---

[2]http://delab.csd.auth.gr/geosocial2/

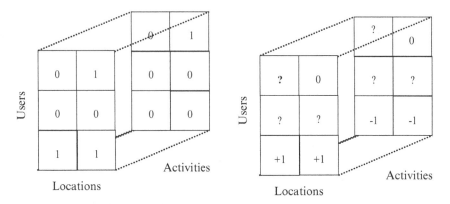

Fig. 6.4 Tensor's data representation of the ITR algorithm (*left*) and the RPCLAF algorithms (*right*)

$$
\theta_{u,l,a,a'} := \begin{cases}
+1, & \text{if } \mathcal{A}_{u,l,a} > \mathcal{A}_{u,l,a'} \mid (u,l,a) \in I_i \wedge (u,l,a') \notin I_i; \\
0, & \text{if } \mathcal{A}_{u,l,a} = \mathcal{A}_{u,l,a'} \mid (u,l,a) \in I_i \wedge (u,l,a') \in I_i; \\
-1, & \text{if } \mathcal{A}_{u,l,a} < \mathcal{A}_{u,l,a'} \mid (u,l,a) \notin I_i \wedge (u,l,a') \in I_i; \\
?, & \text{if } (u,l,a) \notin I_i \vee (u,l,a') \notin I_i
\end{cases} \tag{6.5}
$$

where I_i denotes the location-activity pairwise preferences for user i in tensor $\mathcal{A}$, $\mathcal{A}_{u,l,a}$ denotes the preference of user u on the activity a that she performed in location l, whereas $\mathcal{A}_{u,l,a'}$ denotes the preference of user u on the activity k' that she performed in location l. Based on Eq. (6.5), RPCLAF distinguishes between positive and negative location-activity pairwise preferences and missing values to learn a personalized ranking of activities/locations. The idea is that positive (+1) and negative examples (−1) are only generated from observed location-activity pairwise preferences. Observed location-activity pairwise preferences are interpreted as positive feedback (+1), whereas the non-observed location-activity pairwise preferences are negative (−1) evidences. All other entries are assumed to be either missing (?) or zero values.

To give a more clear view of the tensor representation based on ranking, in Fig. 6.4 we compare the tensor representation of the ITR [15] algorithm, with the tensor representation of the RPCLAF.

The left-hand side of Fig. 6.4 shows the tensor representation of the ITR algorithm [15], where the positive feedback is interpreted as 1 and the rest as 0. The right-hand side of Fig. 6.4 shows the tensor representation of the RPCLAF algorithm where the observed location-activity pairwise preferences are considered positive feedback (+1) while the non-observed location-activity pairwise preferences are marked as negative feedback (−1). All other entries are either missing (?) or zero values. For example, in the right-hand side of Fig. 6.4, the value of tensor element $\mathcal{A}_{3,1,1}$ is +1, because it holds $\mathcal{A}_{3,1,1} > \mathcal{A}_{3,1,2}$, whereas the value of tensor element $\mathcal{A}_{3,1,2} = -1$ because $\mathcal{A}_{3,1,2} < \mathcal{A}_{3,1,1}$.

6.3 Graph-Based Models

6.3.1 FriendSensing Algorithm

Quercia et al. [8, 9] proposed a framework called FriendSensing, which automatically recommends friends by logging and analyzing collocation data. In particular, using short-range technologies (e.g. Bluetooth), mobile phones "sense" and record other mobile devices, which are in geographical proximity. FriendSensing then processes those records and recommends to users people they may know. It measures friendship probability as a function of two variables (i.e. frequency and duration). Frequency keeps track of how many times user A has met user B (i.e., $freq(A, B)$). Duration counts how long a meeting lasted (i.e., $dur(A, B)$). Rather than using absolute frequency and duration values, Friendsensing instead takes their rank. Based on the frequency rank the probability of A befriending B is given by:

$$p(A \rightarrow B) \propto \frac{1}{rankFreq_A(B) + 1}, \qquad (6.6)$$

where:

$$rankFreq_A(B) = |\ \{C : freq(A, C) > freq(A, B)\}\ | + 1.$$

As shown in Eq. (6.6), the probability of A befriending B depends on the number of people that have met A more frequently than B has done. Similarly by replacing frequency with duration, the friendship probability is given by:

$$p(A \rightarrow B) \propto \frac{1}{rankDur_A(B) + 1}, \qquad (6.7)$$

where:

$$rankDur_A(B) = |\ \{C : dur(A, C) > dur(A, B)\}\ | + 1.$$

FriendSensing builds a directed and weighted unipartite social network, where each link weight denotes the aforementioned friendship probabilities. On this network, FriendSensing tested several Markov chain-based algorithms to provide friend recommendations. Specifically, they have compared several Markov Chain-based algorithms (i.e. PageRank with prior, k-Markov Chain, and HITS with prior).

PageRank with prior, for example, reflects the idea that a web page is important if there are many pages linking to it, and if those pages are important themselves. It operates as follows: consider a random walker that starts from node v_x. The random walker chooses randomly among the available edges every time, except that before making a choice, she goes back to node v_x with probability c (restart). Thus, the

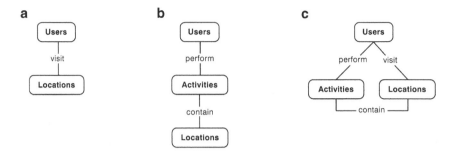

Fig. 6.5 Visual representation of (**a**) the User-Location model, (**b**) the User-Activity and Activity-location Model, and (**c**) the CLM model adapted from [6]

relevance score of node v_x with respect to node v_y is defined as the steady-state probability r_{v_x,v_y} that the random walker will finally stay at node v_y, as shown by Eq. (6.8):

$$\mathbf{r}_{v_x} = c \cdot A \cdot \mathbf{r}_{v_x} + (1 - c) \cdot \mathbf{e}_{v_x} \tag{6.8}$$

where $\mathbf{e}_{v_x}$ is the $n \cdot 1$ starting vector with the v_x-th element equal to 1 and 0 for the other elements of the vector, whereas A is the adjacency matrix of the social network.

6.3.2 CADC Algorithm

The CADC algorithm is part of the Collaborative Location Recommendation (CLR) framework [6]. CLR collects users' GPS trajectory data and represents it with a graph-based structure, denoted as Community Location Model or CLM graph. The Community-based Agglomerative-Divisive Clustering (CADC) algorithm co-clusters the data of the CLM graph into groups of similar users, similar locations and similar activities. In addition, CADC supports incremental updates of the groups when new GPS trajectory data arrives. That is, the new inserted data are incrementally re-clustered without re-clustering the whole CLM graph. Notice also that CLR considers different user classes (i.e. pattern users, normal users and travelers), according to the sequence of locations that they visited in their daily activities. Thus, CLR can provide location recommendations from different classes of users.

As far as the nature of CLM graph is concerned, it is a tripartite graph that captures adequately the relations between users, activities and locations. For example, Fig. 6.5a shows the User-Location model as a bipartite graph, whereas Fig. 6.5b shows the User-Activity and Activity-Location Model as two separate bipartite graphs.

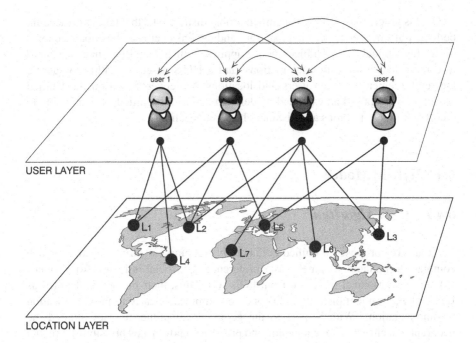

USER LAYER

LOCATION LAYER

Fig. 6.6 Visual representation of the two-layered graph of users and locations

In contrast to the aforementioned models, as shown in Fig. 6.5c, the CLM model exploits: (1) the spatial properties of locations, (2) the temporal-spatial properties of activities, as they are temporal sequences of visited locations, and (3) the long-term spatial properties of users, as they visit different locations due to their long term habits or geographical limits.

6.3.3 ST-Unified Algorithm

Cao et al. [2] introduced a general framework for mining significant semantic locations from GPS data. Semantic locations are considered to their model to leverage the quality of location recommendations. For example, this avoids situations where GPS data with different coordinates (e.g. different geographical latitude and longitude) represent the same semantic location. Also, it avoids cases where a single raw location represents more than one semantic locations. Moreover, they proposed a model that assigns significance to the extracted locations. That is, they model the connections between locations and the connections between location and users using a two layered graph with location-location and user-location components. Figure 6.6 shows the two-layered graph of users and locations.

On this graph, they proposed a unified probabilistic model that takes into account both the links between users and locations and the links between locations, denoted as Unified model. The Unified model applies both a PageRank-like algorithm to capture location-location interactions, and a HITS-based algorithm to capture the reinforcement between users and locations. Moreover, based on the Unified model, they proposed an extended model, denoted as ST-Unified, which is able to incorporate stay durations and distances between locations.

6.4   Hybrid Models

6.4.1   USG Algorithm

Ye et al. [16] proposed a unified collaborative recommendation algorithm, which combines three factors, namely: user preference (U), social influence from friends (S), and geographical influence from POIs (G). Thus, their method is denoted as USG. The aforementioned three factors are combined linearly to provide location recommendations. With respect to the geographical influence, they studied the geographical clustering phenomenon and proposed a power-law probabilistic model to capture the geographical influence among Points of Interest (POIs). The authors evaluated their proposed method over the Foursquare and Whrrl data sets, and discovered among others, that the geographical influence is more important than the social influence and that item similarity is not as accurate as user similarity due to a lack of user check-ins.

6.4.2   SPG Algorithm

Similar to the USG algorithm, the SPG [13] algorithm also combines three features, namely: Social features (S) from the unipartite friend-friend network, Place features (P) from the bipartite place-friend network (i.e. links among users who visit the same places), and Global features (G) that can be computed for any couple of users even if they do not share any friend or place (e.g. the geographic distance between two users). Based on the above features, they use a supervised learning framework, to predict future friendship links among users. In particular, they handle the link prediction problem as a binary classification problem, where the two class labels show the existence or not of a link between two users. They compared the performance of the following classifiers: J48, Naive Bayes, model trees with linear regression on the leaves, and random forests. These classifiers were trained based on the previous features, and predict whether there exists an edge between a pair of nodes or not (two class labels). Experiments have shown that the combination of the three types of features (i.e. Social, Place and Global) can leverage the quality of friend recommendations.

6.4.3 Eventer Algorithm

Eventer [4, 5] allows users of Facebook to rate events that they have attended or have beliefs (preference or not) on them. It uses geo-location techniques to map the user's IP address to her geographical location and crawls concert events from Last.fm. Events are gathered from Last.fm using its web service API (REST). Then, it combines content-based filtering (i.e. artist-artist similarities from Last.fm, event-event feature similarities, etc.) with collaborative filtering (i.e. user ratings on past events) and the social connections of users (i.e. friendship network) to compute user-event similarities. Based on these similarities, it provides personalized event recommendations to users.

6.5 Further Reading

In addition to the methods presented in this chapter, there are many other interesting approaches, which incorporate either additional information sources in their models or tackle other specialized problems in LBSNs.

Sadilek et al. [10] proposed the FLAP (Friendship-Location Analysis and Prediction) system, which provides link and location prediction in LBSNs. For link prediction, FLAP infers social ties by considering patterns in friendship formation, the content of people's messages, and user location. For location prediction, FLAP implements a scalable probabilistic model of human mobility, where they treat users with known GPS positions as noisy sensors of the location of their friends.

Ying et al. [17] proposed Geographic-Textual-Social Based Followee Recommendation (GTS-FR) for recommending interesting followees to users in asymmetrical LBSNs (e.g. Twitter that has no reciprocal connections among users). GTS-FR combines features from three properties in LBSN data, namely: Social Property (SP), Geographical Property (GP), and Textual Property (TP). SP exploits information from the unipartite and directed followee-follower network. GP exploits information from the potential similarity between the trajectories (i.e. trips) of different users. The trajectory reflects the detail of the user's activity. TP exploits information from comments/tags that users write either within their trips or for other users' trips.

Yu et al. [18] proposed an approach that recommends friends to users in LBSNs, by combining GPS information and social network structures to build a heterogeneous information network. Links inside this network reflect both people's geographical information, and their social relationships. Their approach employs a random walk process that runs on the information network to estimate link relevance among users and provide friend recommendation in LBSNs.

Daly et al. [3] identified four categories of events (i.e. Local Not Social, Social and Local, Social Not Local, Global) based on local and social properties, where the location of the event is unknown. Social classification takes into account the social density of the event participants, whereas location classification considers the distance between the attendees of an event. After the different event categories are identified, they recommend events to new uses. They have shown experimentally that if an event is location independent, then its popularity should be considered for recommending it to new users.

Saez-Trumper et al. [11] studied different strategies for recommending to users "potential guests" for inviting them to an event. Their models combine the user preferences with her geographic closeness to an event. They have experimentally shown that simple models such as Bayesian or linear regression, which exploit the above features, can provide accurate recommendations.

References

1. B. Betim, S. Thorsten, A recommendation system for spots in location-based online social networks, in *Proceedings of the 4th Workshop on Social Network Systems (SNS)*, Salzburg (2011), pp. 4:1–4:6
2. X. Cao, G. Cong, C. Jensen, Mining significant semantic locations from GPS data. Proc. VLDB Endowment **3**(1–2), 1009–1020 (2010)
3. E.M. Daly, W. Geyer, Effective event discovery: using location and social information for scoping event recommendations, in *Proceedings of the Fifth ACM conference on Recommender Systems* (ACM, New York, 2011), pp. 277–280
4. M. Kayaalp, T. Ozyer, S.T. Ozyer, A collaborative and content based event recommendation system integrated with data collection scrapers and services at a social networking site, in *Proceedings of the International Conference on Advances in Social Network Analysis and Mining (ASONAM)*, Athens (2009), pp. 113–118
5. M. Kayaalp, T. Ozyer, S.T. Ozyer,   A mash-up application utilizing hybridized filtering techniques for recommending events at a social networking site. Soc. Netw. Anal. Min. **1**(3), 231–239 (2011)
6. K.W.T. Leung, D.L. Lee, W.C. Lee, CLR: a collaborative location recommendation framework based on co-clustering, in *Proceedings of the 34th ACM SIGIR International Conference on Research and Development in Information Retrieval (SIGIR)*, Beijing (2011), pp. 305–314
7. A. Papadimitriou, P. Symeonidis, Y. Manolopoulos,   Geo-social recommendations,   in *Proceedings of the RecSys Workshop on Personalization on Mobile Applications (PeMA)*, Chicago, IL (2011)
8. D. Quercia, L. Capra,   Friendsensing: recommending friends using mobile phones,   in *Proceedings of the 3rd ACM Conference on Recommender Systems (RecSys)*, New York, NY (2009), pp. 273–276
9. D. Quercia, J. Ellis, L. Capra, Using mobile phones to nurture social networks. IEEE Pervasive Comput. **9**(3), 12–20 (2010)
10. A. Sadilek, H. Kautz, J.P. Bigham, Finding your friends and following them to where you are, in *Proceedings of the Fifth ACM International Conference on Web Search and Data Mining (WSDM 2012)* (ACM, New York, 2012), pp. 723–732
11. D. Saez-Trumper, D. Quercia, J. Crowcroft, Ads and the city: considering geographic distance goes a long way, in *Proceedings of the Sixth ACM Conference on Recommender Systems* (ACM, New York, 2012), pp. 187–194
12. M. Sattari, M. Manguoglu, I.H. Toroslu, P. Symeonidis, P. Senkul, Y. Manolopoulos, Geo-activity recommendations by using improved feature combination, in *Proceedings of the ACM UbiComp International Workshop on Location-Based Social Networks (LBSN)*, Pittsburgh, PA (2012), pp. 996–1003

13. S. Scellato, A. Noulas, C. Mascolo, Exploiting place features in link prediction on location-based social networks, in *Proceedings of the 17th ACM SIGKDD International Conference on Knowledge Discovery and Data Mining (KDD)*, San Diego, CA (2011), pp. 1046–1054

14. A.P. Singh, G.J. Gordon, Relational learning via collective matrix factorization, in *Proceeding of the 14th ACM SIGKDD International Conference on Knowledge Discovery and Data Mining (KDD)*, Las Vegas, NV (2008), pp. 650–658

15. P. Symeonidis, A. Papadimitriou, Y. Manolopoulos, P. Senkul, I. Toroslu, Geo-social recommendations based on incremental tensor reduction and local path traversal, in *Proceedings of the 3rd ACM SIGSPATIAL International Workshop on Location-Based Social Networks (LBSN)*, Chicago, IL (2011), pp. 89–96

16. M. Ye, P. Yin, W.C. Lee, D.L. Lee, Exploiting geographical influence for collaborative point-of-interest recommendation, in *Proceedings of the 34th ACM SIGIR International Conference on Research and Development in Information Retrieval (SIGIR)*, Beijing (2011), pp. 325–334

17. J.J. Ying, E.H. Lu, V.S. Tseng, Followee recommendation in asymmetrical location-based social networks, in *Proceedings of the 2012 ACM Conference on Ubiquitous Computing* (ACM, New York, 2012), pp. 988–995

18. X. Yu, A. Pan, L.-A. Tang, Z. Li, J. Han, Geo-friends recommendation in GPS-based cyber-physical social network, in *IEEE International Conference on Advances in Social Networks Analysis and Mining (ASONAM)* (IEEE, Kaohsiung, Taiwan 2011), pp. 361–368

19. V. Zheng, B. Cao, Y. Zheng, X. Xie, Q. Yang, Collaborative filtering meets mobile recommendation: a user-centered approach, in *Proceedings of the 24th AAAI Conference on Artificial Intelligence (AAAI)*, Atlanta, GA (2010)

20. V. Zheng, Y. Zheng, X. Xie, Q. Yang, Collaborative location and activity recommendations with GPS history data, in *Proceedings of the 19th International Conference on World Wide Web (WWW)*, New York, NY (2010), pp. 1029–1038

21. Y. Zheng, X. Xiem, W.Y. Ma, Geolife: a collaborative social networking service among user, location and trajectory. IEEE Data Eng. Bull. **33**(2), 32–39 (2010)

22. V. Zheng, Y. Zheng, X. Xie, Q. Yang, Towards mobile intelligence: learning from GPS history data for collaborative recommendation. Artif. Intell. **184–185**, 17–37 (2012)

Chapter 7
Comparison

This chapter compares and categorizes the algorithms described in Chap. 6 on the basis of their characteristics. We categorized them based on: (1) the kind of recommendation they provide (i.e., generic or personalized), (2) the type of recommendation they provide (i.e. friend, location, activity, and event), (3) the data representation they use for their model (i.e. matrix, tensor, graph), (4) the technique they are based on (i.e. probabilistic, semantic, collaborative filtering, etc.), and (5) the data sets and the metrics they use in their experiments.

7.1 Generic vs. Personalized Recommendations

Table 7.1 presents the categorization of algorithms in terms of generic vs. personalized recommendations. As expected, most methods try to provide personalized recommendations. Notice that, this is the advantage of a recommender system over a search engine. That is, the former brings personalized results for each user, whereas a search engine provides just a generic ranking of the results.

7.2 Recommendation Type

Table 7.2 presents the categorization of algorithms in terms of recommendations types (i.e. friend, location, activity, and event). As shown, most methods focus on providing either location or activity recommendations. However, there are not many methods that can provide more than two types of recommendations. In particular, ITR is the only method providing three different types of recommendations. Finally, event recommendation, which is a specialized category of activity recommendation, deserves further research attention.

P. Symeonidis et al., *Recommender Systems for Location-based Social Networks*,   81
SpringerBriefs in Electrical and Computer Engineering,
DOI 10.1007/978-1-4939-0286-6_7, © The Author(s) 2014

Table 7.1 Algorithms vs.
generic/personalized
recommendations

Algorithm	Generic	Personalized
CLAF [11]	✓	
IFC [6]	✓	
RMF [1]		✓
PCLAF [10]		✓
ITR [8]		✓
RPCLAF [12]		✓
FriendSensing [5]		✓
CADC [4]		✓
ST-Unified [2]	✓	
USG [9]		✓
SPG [7]		✓
Eventer [3]		✓

Table 7.2 Algorithms vs. recommendation types

Algorithm	Friend	Location	Activity	Event
CLAF [11]		✓	✓	
IFC [6]		✓	✓	
RMF [1]		✓		
PCLAF [10]		✓	✓	
ITR [8]	✓	✓	✓	
RPCLAF [12]		✓	✓	
FriendSensing [5]	✓			
CADC [4]		✓		
ST-Unified [2]		✓		
USG [9]		✓		
SPG [7]	✓			
Eventer [3]				✓

7.3   Data Representation

Table 7.3 presents the categorization of algorithms in terms of data representation. As shown, there is a balanced allocation of methods into the three specified categories (matrix, tensor, graph). Notice that a tensor is a higher order matrix, that captures more dimensions of the problem (i.e. users, locations, activities, ratings, tags, etc.) incorporating richer information into the model. Regarding the graph representations, they can also incorporate richer information (i.e. tripartite graphs).

7.4   Categories of Problem Modeling

Table 7.4 presents the problem modeling categories in LBSNs.

Table 7.3 Data representation

Algorithm	Matrix	Tensor	Graph
CLAF [11]	✓		
IFC [6]	✓		
RMF [1]	✓		
PCLAF [10]		✓	
ITR [8]		✓	
RPCLAF [12]		✓	
FriendSensing [5]			✓
CADC [4]			✓
ST-Unified [2]			✓
USG [9]	✓		
SPG [7]			✓
Eventer [3]	✓		

Table 7.4 Categories of recommendation algorithms

Algorithm	Collab. filtering	Factorization	Graph-based	Hybrid	Semantic	Probabilistic	Classification	Clustering
CLAF [11]	✓	✓						
IFC [6]	✓	✓						
RMF [1]	✓	✓						
PCLAF [10]	✓	✓						
ITR [8]	✓	✓						
RPCLAF [12]	✓	✓						
FriendSensing [5]			✓					
CADC [4]								✓
ST-Unified [2]			✓	✓	✓			
USG [9]	✓			✓		✓		
SPG [7]				✓			✓	
Eventer [3]	✓			✓				

A method can incorporate more than one techniques for modeling the recommendation problem in LBSNs. For example, SPG [7] proposed a *hybrid* supervised learning approach to link prediction, modeling it as a binary classification problem. As also shown in Table 7.4, the mainstream of problem modeling examines methods that combine collaborative filtering with matrix/tensor factorization. This is reasonable, since collaborative filtering is a very popular technique in recommender systems for its merits. That is, it provides accurate recommendations with serendipity and diversity, based on the "word of mouth" practice.

7.5   Algorithms vs. Data Sets

Table 7.5 presents the data sets that have been used for the experimental evaluation of the selected algorithms. As shown, the most popular data set is Geolife,[1] which has been used for testing by CLAF, PCLAF, RPCLAF, IFC and ITR algorithms. The GeoSocial[2] data set has been used for testing of ITR algorithm. The Reality Mining[3] data set is used by FriendSensing. The Gowalla[4] data set has been used in RMF and SPG algorithms. Notice that the Gowalla web site is bought out by facebook and it is now shut down. Finally, there are also crawled data sets from Foursquare and Facebook web sites.

7.6   Algorithms vs. Metrics

Table 7.6 presents the metrics that have been used for the experimental evaluation of the selected algorithms. As shown, Precision-Recall, nDCG and RMSE have

Table 7.5 Algorithms vs. data sets

Algorithm	Geolife	GeoSocialRec	Gowalla	Reality mining	Foursquare and Whrrl	Facebook and Last.fm	Other
CLAF [11]	✓						
IFC [6]	✓						
RMF [1]			✓				
PCLAF [10]	✓						
ITR [8]	✓	✓					
RPCLAF [12]	✓						
FriendSensing [5]				✓			
CADC [4]							✓
ST-Unified [2]							✓
USG [9]					✓		
SPG [7]			✓				
Eventer [3]						✓	

---

[1] http://www.cse.ust.hk/~vincentz/aaai10.uclaf.data.mat

[2] http://delab.csd.auth.gr/geosocial

[3] http://realitycommons.media.mit.edu/realitymining.html

[4] http://en.wikipedia.org/wiki/Gowalla

Table 7.6 Algorithms vs. metrics

Algorithm	nDCG	Precision-Recall	RMSE	MAE	AUC	MAP	False positive - FN
CLAF [11]	✓						
IFC [6]			✓	✓			
RMF [1]		✓	✓				
PCLAF [10]	✓		✓				
ITR [8]		✓					
RPCLAF [12]			✓		✓		
FriendSensing [5]		✓					
CADC [4]	✓						
ST-Unified [2]	✓	✓				✓	
USG [9]		✓					
SPG [7]		✓			✓		
Eventer [3]							✓

been used by many methods. At this point, it is mentionable that none of the methods under investigation has used the same experimental protocol, data sets and metrics to compare with other works. Thus, we will not report any result in terms of recommendation and prediction accuracy. In any case, the creation of a common experimental protocol with a common data set, should be one of the primary considerations in future works.

References

1. B. Betim, S. Thorsten, A recommendation system for spots in location-based online social networks, in *Proceedings of the 4th Workshop on Social Network Systems (SNS)*, Salzburg (2011), pp. 4:1–4:6
2. X. Cao, G. Cong, C. Jensen, Mining significant semantic locations from GPS data. Proc. VLDB Endowment **3**(1–2), 1009–1020 (2010)
3. M. Kayaalp, T. Ozyer, S.T. Ozyer, A collaborative and content based event recommendation system integrated with data collection scrapers and services at a social networking site, in *Proceedings of the International Conference on Advances in Social Network Analysis and Mining (ASONAM)*, Athens (2009), pp. 113–118
4. K.W.T. Leung, D.L. Lee, W.C. Lee, CLR: a collaborative location recommendation framework based on co-clustering, in *Proceedings of the 34th ACM SIGIR International Conference on Research and Development in Information Retrieval (SIGIR)*, Beijing (2011), pp. 305–314
5. D. Quercia, L. Capra, Friendsensing: recommending friends using mobile phones, in *Proceedings of the 3rd ACM Conference on Recommender Systems (RecSys)*, New York (2009), pp. 273–276

6. M. Sattari, M. Manguoglu, I.H. Toroslu, P. Symeonidis, P. Senkul, Y. Manolopoulos, Geo-activity recommendations by using improved feature combination, in *Proceedings of the ACM UbiComp International Workshop on Location-Based Social Networks (LBSN)*, Pittsburgh, PA (2012), pp. 996–1003

7. S. Scellato, A. Noulas, C. Mascolo, Exploiting place features in link prediction on location-based social networks, in *Proceedings of the 17th ACM SIGKDD International Conference on Knowledge Discovery and Data Mining (KDD)*, San Diego, CA (2011), pp. 1046–1054

8. P. Symeonidis, A. Papadimitriou, Y. Manolopoulos, P. Senkul, I. Toroslu, Geo-social recommendations based on incremental tensor reduction and local path traversal, in *Proceedings of the 3rd ACM SIGSPATIAL International Workshop on Location-Based Social Networks (LBSN)*, Chicago, IL (2011), pp. 89–96

9. M. Ye, P. Yin, W.C. Lee, D.L. Lee, Exploiting geographical influence for collaborative point-of-interest recommendation, in *Proceedings of the 34th ACM SIGIR International Conference on Research and Development in Information Retrieval (SIGIR)*, Beijing (2011), pp. 325–334

10. V. Zheng, B. Cao, Y. Zheng, X. Xie, Q. Yang, Collaborative filtering meets mobile recommendation: a user-centered approach, in *Proceedings of the 24th AAAI Conference on Artificial Intelligence (AAAI)*, Atlanta, GA (2010)

11. V. Zheng, Y. Zheng, X. Xie, Q. Yang, Collaborative location and activity recommendations with GPS history data, in *Proceedings of the 19th International Conference on World Wide Web (WWW)*, New York (2010), pp. 1029–1038

12. V. Zheng, Y. Zheng, X. Xie, Q. Yang, Towards mobile intelligence: learning from GPS history data for collaborative recommendation. Artif. Intell. **184–185**, 17–37 (2012)

Part III
Implementing a Real-World LBSN

Chapter 8
Real Geo-Social Recommender System

This chapter presents a real-world recommender system for LBSNs. GeoSocialRec allows to test, evaluate and compare different recommendation styles in an online setting, where the users of GeoSocialRec actually receive recommendations during their check-in process.

8.1 Geosocial System Description

The GeoSocialRec recommender system consists of several components. The system's architecture is illustrated in Fig. 8.1, where three main sub-systems are described: (1) the Web Site, (2) the Database Profiles and (3) the Recommendation Engine. In the following sections, we describe each sub-system of GeoSocialRec in detail.

8.1.1 GeoSocialRec Web Site

The GeoSocialRec system uses a web site[1] to interact with the users. The web site consists of four subsystems: (1) the friend recommendation, (2) the location recommendation, (3) the activity recommendation, and (4) the check-in subsystem. The friend recommendation subsystem is responsible for evaluating incoming data from the Recommendation Engine of GeoSocialRec and providing updated friend recommendations. To provide such recommendations, the web site subsystem implements the FriendLink algorithm [4] and also considers the geographical distance between users and check-in points. The same applies to the location and

---

[1]http://delab.csd.auth.gr/geosocialrec

P. Symeonidis et al., *Recommender Systems for Location-based Social Networks*,   89
SpringerBriefs in Electrical and Computer Engineering,
DOI 10.1007/978-1-4939-0286-6_8, © The Author(s) 2014

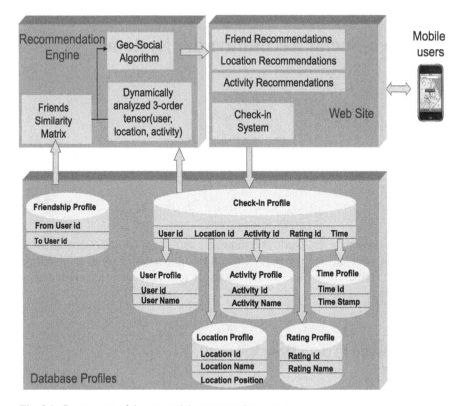

Fig. 8.1 Components of the geo-social recommender system

activity recommendation sub-systems, where new and updated location and activity recommendations are presented to the user as new check-ins are stored in the Database profiles. Finally, the check-in subsystem is responsible for passing the data inserted by the users to the respective Database profiles.

Figure 8.2 presents a scenario where the GeoSocialRec system recommends four possible friends to user *Panagiotis Symeonidis*. As shown, the first table recommends Anastasia Kalou and Ioanna Kontaki, who are connected to him with 2-hop paths. The results are ordered based on the second to last column of the table, which indicates the number of common friends that the target user shares with each possible friend. As shown in Fig. 8.2, Anastasia Kalou is the top recommendation because she shares three common friends with the target user. The common friends are then presented in the last column of the table. The second table contains two users, namely Manolis Daskalakis and George Tsalikidis, who are connected to the target user via 3-hop paths. The last column of the second table indicates the number of found paths that connect the target user with the recommended friends. Manolis Daskalakis is now the top recommendation, because he is connected to Panagiotis

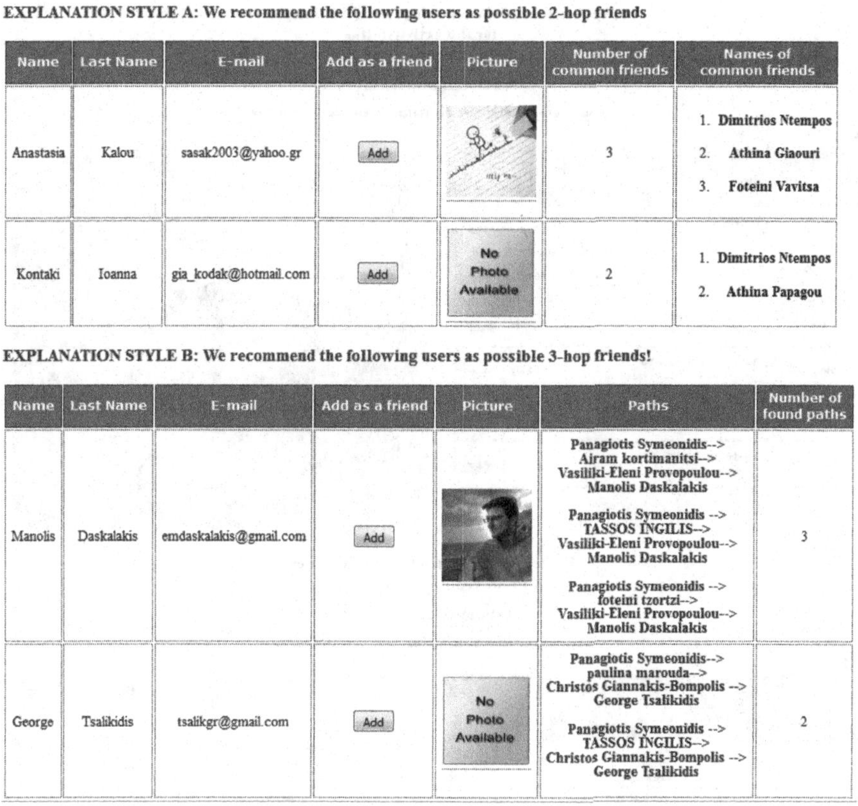

EXPLANATION STYLE A: We recommend the following users as possible 2-hop friends

Name	Last Name	E-mail	Add as a friend	Picture	Number of common friends	Names of common friends
Anastasia	Kalou	sasak2003@yahoo.gr	Add		3	1. **Dimitrios Ntempos** 2. **Athina Giaouri** 3. **Foteini Vavitsa**
Kontaki	Ioanna	gia_kodak@hotmail.com	Add	No Photo Available	2	1. **Dimitrios Ntempos** 2. **Athina Papagou**

EXPLANATION STYLE B: We recommend the following users as possible 3-hop friends!

Name	Last Name	E-mail	Add as a friend	Picture	Paths	Number of found paths
Manolis	Daskalakis	emdaskalakis@gmail.com	Add		**Panagiotis Symeonidis--> Airam kortimanitsi--> Vasiliki-Eleni Provopoulou--> Manolis Daskalakis** **Panagiotis Symeonidis --> TASSOS INGILIS--> Vasiliki-Eleni Provopoulou--> Manolis Daskalakis** **Panagiotis Symeonidis --> foteini tzortzi--> Vasiliki-Eleni Provopoulou--> Manolis Daskalakis**	3
George	Tsalikidis	tsalikgr@gmail.com	Add	No Photo Available	**Panagiotis Symeonidis--> paulina marouda--> Christos Giannakis-Bompolis --> George Tsalikidis** **Panagiotis Symeonidis --> TASSOS INGILIS--> Christos Giannakis-Bompolis --> George Tsalikidis**	2

Fig. 8.2 Friend recommendations provided by the GeoSocialRec system

Symeonidis via three 3-hop paths. It is obvious that the second explanation style is more analytical and detailed, since users can see, in a transparent way, the paths that connect them with the recommended friends.

Figure 8.3a shows a location recommendation, while Fig. 8.3b depicts an activity recommendation. As shown in Fig. 8.3a, the target user can provide to the system the activity she wants to do and the place she is (i.e. Bar in Athens). Then, the system provides a map with bar places (i.e. place A, place B, place C, etc.) along with a table, where these places are ranked based on the number of users' check-ins and their average rating. As shown in Fig. 8.3a, the top recommended Bar is Mojo (i.e. place A), which is visited three times (from the target user's friends) and is rated highly (i.e. 5 stars). Regarding the activity recommendation, as shown in Fig. 8.3b, the user selects a nearby city (i.e. Thessaloniki) and the system provides activities that she could perform. In this case, the top recommended activity is sightseeing the White Tower of Thessaloniki, because it is visited 14 times and has an average rating of 4.36.

We recommend the following Point(s) of Interest
for the Activity: Bar
in the city of:Athens

EXPLANATION STYLE A:

We recommend the following POI's (Point of Interest) based on total Check-ins!

a

A/A	Point Of Interest	POI Address	Explanation Style A: Total Check-Ins	Average Rating from style A	Go To
A	Mojo	Παπαδιαμαντοπούλου 8, Ζωγράφου 157 71, Ελλάδα	3	5.0000	Move!
B	A for Athens	Ερμού 82, Αθήνα 105 55, Ελλάδα	3	3.6667	Move!
C	Rox Box	Ειρήνης 2-10, Φιλαδέλφεια Χαλκηδόνα 143 41, Ελλάδα	2	4.5000	Move!
D	Holy Spirit	Λαοδίκης 18, Γλυφάδα 166 74, Ελλάδα	2	4.0000	Move!
E	Mo Better	Κωλέττη 28-42, Αθήνα, Ελλάδα	2	4.0000	Move!
F	Allo Bar	Thoukydidou 9-13, Chalandri 15232, Greece	2	2.0000	Move!

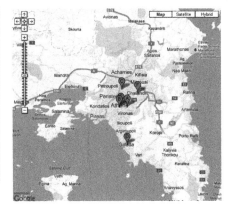

We recommend the following Activities
in the city of:Thessaloniki

EXPLANATION STYLE A:

We recommend the following activities based on total Check-ins!

b

A/A	Activity	Point Of Interest	POI Address	Explanation Style A: Total Check-Ins	Average Rating from style A	Go To
A	Sight-seeing	White Tower	Nikis Avenue-- Paralia Thessalonikis	14	4.3571	Move!
B	Education	Aristotle University of Thessaloniki	Egnatia & Kondriktonos-- Aristotle Campus	13	4.2308	Move!
C	Sight-seeing	Aristotelous Square	Aristotle Square- -City Center	11	4.1818	Move!
D	Museums	Archaeological Museum of Thessaloniki	Ανδρόνικου, Θεσσαλονίκη 54621, Ελλάς--	8	4.0000	Move!
E	Cinema	Video Land	Ιθάκης 63-71, Εύοσμο 56224, Ελλάς--	8	2.6250	Move!
F	Bar	BRISTOL	George Papandreou 24-- Poseidonio	7	4.2857	Move!

Fig. 8.3 The geo-social recommender system provides (**a**) Point of Interest or Location Recommendations, and (**b**) Activity recommendations

8.1.2 GeoSocialRec Database Profiles

The database that supports the GeoSocialRec system is a MySQL(v.5.5.8)[2] database. MySQL is an established Database Management System (DBMS), which is widely used in on-line, dynamic, database driven websites.

The database profile sub-system contains five profiles where data about the users, locations, activities and their corresponding ratings are stored. As shown in Fig. 8.1, this data are received by the Check-In profile and along with the Friendship profile, they provide the input for the Recommendation Engine sub-system. Each table field represents the respective data that is collected by the Check-In profile. User-id, Location-id and Activity-id refer to specific ids given to users, locations and activities respectively.

8.1.3 GeoSocialRec Recommendation Engine

The recommendation engine is responsible for collecting the data from the database and producing the recommendations, which will then be displayed on the web site. As shown in Fig. 8.1, the recommendation engine constructs a friends similarity matrix by implementing the FriendLink algorithm proposed in [4]. The average geographical distances between users' check-ins are used as link weights. To obtain the weights, we calculate the average distance between all pairs of POIs that two users have checked-in. The recommendation engine also produces a dynamically analyzed three-order tensor, which is firstly constructed by the HOSVD algorithm and is then updated using incremental methods [1, 5], both of which are explained in later sections.

8.2 The FriendLink Algorithm

In this section, we describe the FriendLink [4] algorithm, which can be used for the task of friend recommendations. Here, we describe how FriendLink is applied on GSNs and how the recommendation of friends is performed according to the detected associations.

When using an GSN, users explicitly declare their friends so that they are able to share information location with them (i.e. photos etc.). After some time, the geo-social network accumulates a set of connection data (graph of friendships), which can be represented by an undirected graph.

---

[2]http://www.mysql.com

FriendLink assumes that persons in an GSN can use all the pathways connecting them, proportionally to the pathway lengths. Thus, two persons who are connected with many unique pathways have a high possibility to know each other, proportionally to the length of the pathways they are connected with.

Definition 1. The similarity $sim(v_x, v_y)$ between two graph nodes v_x and v_y is defined as the counts of paths of varying length ℓ from v_x to v_y:

$$sim(v_x, v_y) = \sum_{i=2}^{\ell} \frac{1}{i-1} \cdot \frac{\left| paths_{v_x,v_y}^i \right|}{\prod_{j=2}^{i}(n-j)} \tag{8.1}$$

where:

- n is the number of vertices in a graph G,
- ℓ is the maximum length of a path between the graph nodes v_x and v_y (excluding paths with cycles). By the term "paths with cycles" we mean that a path can not be closed (cyclic). Thus, a node can appear once in a path (e.g. path $v_1 \rightarrow v_2 \rightarrow v_3 \rightarrow v_1 \rightarrow v_5$ is not acceptable because v_1 is traversed twice),
- $\frac{1}{i-1}$ is an "attenuation" factor that weights paths according to their length ℓ. Thus, a two-step path measures the non-attenuation of a link with value equals to 1 ($\frac{1}{2-1} = 1$). A three-step path measures the attenuation of a link with value equals to $\frac{1}{2}$ ($\frac{1}{3-1} = \frac{1}{2}$) etc. In this sense, we use appropriate weights to allow the lower effectiveness of longer path chains.
- $\left| paths_{v_x,v_y}^{\ell} \right|$ is the count of all length-ℓ paths from v_x to v_y,
- $\prod_{j=2}^{i}(n-j)$ is the count of all possible length-ℓ paths from v_x to v_y, if each vertex in graph G was linked with all other vertices. By using the fraction $\dfrac{\left| paths_{v_x,v_y}^{\ell} \right|}{\prod_{j=2}^{i}(n-j)}$, the similarity measure is normalized and takes values in [0,1]. If two nodes are similar, then we expect the value $sim(v_x, v_y)$ to be close to 1. On the other hand, if the two nodes are dissimilar, we expect the value $sim(v_i, v_j)$ to be close to 0.

FriendLink finds similarities between nodes in an undirected graph constructed from these connection data. The FriendLink algorithm uses as input the connections of a graph $\mathcal{G}$ and outputs a similarity matrix between any two nodes in $\mathcal{G}$. Therefore, friends can be recommended to a target user u according to their weights in the similarity matrix.

8.3 The Incremental Tensor Reduction Approach

In this section we provide details on how HOSVD is applied on tensors and how location/activity recommendation is performed based on the detected latent associations.

The Incremental Tensor Reduction (ITR) approach initially constructs a tensor, based on usage data triplets $\{u, l, a\}$ of users, location and activity. The motivation is to use all three objects that interact inside a location-based social network. Consequently, we proceed to the unfolding of $\mathcal{A}$, where we build three new matrices. Then, we apply SVD in each new matrix. Finally, we build the core tensor $\mathcal{S}$ and the resulting tensor $\hat{\mathcal{A}}$. The six steps of the proposed approach are summarized as follows:

- *Step 1*: The initial tensor $\mathcal{A}$ construction, which is based on usage data triplets (user, location, activity).
- *Step 2*: The matrix unfoldings of tensor $\mathcal{A}$, where we matricize the tensor in all three modes, creating three new matrices (one for each mode).
- *Step 3*: The application of SVD in all three new matrices, where we keep the c-most important singular values for each matrix.
- *Step 4*: The construction of the core tensor $\mathcal{S}$, that reduces the dimensionality.
- *Step 5*: The construction of the $\hat{\mathcal{A}}$ tensor, that is an approximation of tensor $\mathcal{A}$.
- *Step 6*: Based on the weights of the elements of the reconstructed tensor $\hat{\mathcal{A}}$, we recommend location/activity to the target user u.

Steps 1–5 build a model and can be performed off-line. The recommendation in Step 5 is performed on-line, i.e., each time we have to recommend a location/activity to a user, based on the built model. In the following, we provide more details on each step.

8.3.1 The Initial Construction of Tensor $\mathcal{A}$

From the usage data triplets (user, location, activity), we construct an initial three-order tensor $\mathcal{A} \in R^{I_u \times I_l \times I_a}$, where I_u, I_l, I_a are the numbers of users, locations and activities, respectively. Each tensor element measures the number of times that a user u checked in a location l and made an activity a.

8.3.2 Matrix Unfolding of Tensor $\mathcal{A}$

As described, a tensor $\mathcal{A}$ can be unfolded (matricized), i.e., we build matrix representations of tensor $\mathcal{A}$ in which all the column (row) vectors are stacked one after the other. The initial tensor $\mathcal{A}$ is matricized in all three modes. Thus, after the unfolding of tensor $\mathcal{A}$ for all three modes, we create three new matrices A_1, A_2, A_3, as follows:

$$A_1 \in R^{I_u \times I_l I_a},$$

$$A_2 \in R^{I_l \times I_u I_a},$$

$$A_3 \in R^{I_u I_l \times I_a}$$

8.3.3 Application of SVD on Each Matrix

We apply SVD on the three matrix unfoldings A_1, A_2, A_3. We result, in total, to nine new matrices.

$$A_1 = U^{(1)} \cdot S_1 \cdot V_1^T \tag{8.2}$$

$$A_2 = U^{(2)} \cdot S_2 \cdot V_2^T \tag{8.3}$$

$$A_3 = U^{(3)} \cdot S_3 \cdot V_3^T \tag{8.4}$$

For Tensor Dimensionality Reduction, there are three dimensional parameters to be determined. The numbers c_1, c_2 and c_3 of left singular vectors of matrices $U^{(1)}$, $U^{(2)}$, $U^{(3)}$, respectively, that are preserved. They will determine the final dimensionality of the core tensor S. Since each of the three diagonal singular matrices S_1, S_2 and S_3 are calculated by applying SVD on matrices A_1, A_2 and A_3, respectively, we use different c_1, c_2 and c_3 numbers of principal components for each matrix $U^{(1)}$, $U^{(2)}$, $U^{(3)}$. The numbers c_1, c_2 and c_3 of singular vectors are chosen by preserving a percentage of information of the original S_1, S_2, S_3 matrices after appropriate tuning (the default percentage is set to 50 % of the original matrix).

8.3.4 The Construction of the Core Tensor S

The core tensor S governs the interactions among users, locations and activities. Since we have selected the dimensions of $U^{(1)}$, $U^{(2)}$ and $U^{(3)}$ matrices, we proceed to the construction of S, as follows:

$$S = \mathcal{A} \times_1 U_{c_1}^{(1)^T} \times_2 U_{c_2}^{(2)^T} \times_3 U_{c_3}^{(3)^T}, \tag{8.5}$$

where $\mathcal{A}$ is the initial tensor, $U_{c_1}^{(1)^T}$ is the transpose of the c_1-dimensionally reduced $U^{(1)}$ matrix, $U_{c_2}^{(2)^T}$ is the transpose of the c_2-dimensionally reduced $U^{(2)}$ matrix, $U_{c_3}^{(3)^T}$ is the transpose of the c_3-dimensionally reduced $U^{(3)}$ matrix.

8.3.5 The Construction of Tensor $\hat{\mathcal{A}}$

Finally, tensor $\hat{\mathcal{A}}$ is build as the product of the core tensor $\mathcal{S}$ and the mode products of the three matrices $U^{(1)}$, $U^{(2)}$ and $U^{(3)}$ as follows:

$$\hat{\mathcal{A}} = \mathcal{S} \times_1 U_{c_1}{}^{(1)} \times_2 U_{c_2}{}^{(2)} \times_3 U_{c_3}{}^{(3)}, \tag{8.6}$$

$\mathcal{S}$ is the reduced core tensor, $U_{c_1}{}^{(1)}$ is the c_1-dimensionally reduced $U^{(1)}$ matrix, $U_{c_2}{}^{(2)}$ is the c_2-dimensionally reduced $U^{(2)}$ matrix, $U_{c_3}{}^{(3)}$ is the c_3-dimensionally reduced $U^{(3)}$ matrix.

8.3.6 The Generation of the Location/Activity Recommendation List

Tensor $\hat{\mathcal{A}}$ measures the associations among the users, locations and activities and acts as a model that is used during the recommendation.

Each element of $\hat{\mathcal{A}}$ represents a quadruplet $\{u, l, a, p\}$, where p is the likeliness that user u will visit location l and perform activity a. Therefore, locations/activities can be recommended to u according to their weights associated with $\{u, a\}$ and $\{u, l\}$ pairs, respectively. If we want to recommend N activities to user u for location l, then we select the N corresponding activities with the highest weights.

8.3.7 Inserting New Users, Locations, or Activities Over Time

As new users, locations, or activities are being introduced to the system, the $\hat{\mathcal{A}}$ tensor, which provides the recommendations, has to be updated. The most demanding operation for this task is updating the SVD of the corresponding unfoldings. We can avoid the costly batch recomputation of the corresponding SVD, by considering incremental solutions [1, 5]. Depending on the size of the update (i.e., number of new users, locations, or activities), different techniques have been followed in related research. For small update sizes we can consider the *folding-in* technique [5], whereas for larger update sizes we can consider Incremental SVD techniques [1].

8.3.7.1 Update by Incremental SVD

Folding-in incrementally updates SVD, but the resulting model is not a perfect SVD model, because the space is not orthogonal [5]. When the update size is not big, loss of orthogonality may not be a severe problem in practice. Nevertheless, for larger update sizes the loss of orthogonality may result to an inaccurate SVD model. In this case, we need to incrementally update SVD so as to ensure orthogonality. This can be attained in several ways. Next we describe how to use the approach proposed by Brand [1].

Let $M_{p \times q}$ be a matrix, upon which we apply SVD and maintain the first r singular values, i.e.,

$$M_{p \times q} = U_{p \times r} S_{r \times r} V_{r \times q}^T$$

Assume that each column of matrix $C_{p \times c}$ contains the additional elements. Let $L = U \backslash C = U^T C$ be the projection of C onto the orthogonal basis of U. Let also $H = (I - UU^T)C = C - UL$ be the component of C orthogonal to the subspace spanned by U (I is the identity matrix). Finally, let J be an orthogonal basis of H and let $K = J \backslash H = J^T H$ be the projection of C onto the subspace orthogonal to U. Consider the following identity:

$$[U \; J] \begin{bmatrix} S & L \\ 0 & K \end{bmatrix} \begin{bmatrix} V & 0 \\ 0 & I \end{bmatrix}^T =$$

$$[U(I - UU^T)C/K] \begin{bmatrix} S & U^T C \\ 0 & K \end{bmatrix} \begin{bmatrix} V & 0 \\ 0 & I \end{bmatrix}^T =$$

$$[USV^T \; C] = [M \; C]$$

Like an SVD, the left and right matrixes in the product are unitary and orthogonal. The middle matrix, denoted as Q, is diagonal. To incrementally update the SVD, Q must be diagonalized. If we apply SVD on Q we get:

$$Q = U'S'(V')^T$$

Additionally, define U'', S'', V'' as follows:

$$U'' = [U \; J]U', \; S'' = S', \; V'' = \begin{bmatrix} V & 0 \\ 0 & I \end{bmatrix} V'$$

Then, the updated SVD of matrix $[M \; C]$ is:

$$[M \; C] = [USV^T \; C] = U''S''(V'')^T$$

This incremental update procedure takes $O((p + q)r^2 + pc^2)$ time [2].

Returning to the application of incremental update for new users, locations, or activities, in each case we result with a number of new rows that are appended at the end of the unfolded matrix of the corresponding mode. Therefore, we need an incremental SVD procedure in case we add new rows, whereas the aforementioned method works in case we add new columns. In this case we simply swap U for V and U'' for V''.

8.4 Experimental Configuration

In this section, we study the performance of FriendLink and ITR approaches in terms of friend, location and activity recommendations. To evaluate the aforementioned recommendations we have chosen two real data sets. The first one, denoted as GeoSocialRec data set, is extracted from the GeoSocialRec site.[3] It consists of 102 users, 46 locations and 18 activities. The second data set, denoted as UCLAF [6], consists of 164 users, 168 locations and 5 different types of activities, including "Food and Drink", "Shopping", "Movies and Shows", "Sports and Exercise", and "Tourism and Amusement".

The numbers c_1, c_2, and c_3 of left singular vectors of matrices $U^{(1)}$, $U^{(2)}$, $U^{(3)}$ for ITR, after appropriate tuning, are set to 25, 12 and 8 for the GeoSocialRec dataset, and to 40, 35, 5 for the UCLAF data set. Due to lack of space we do not present experiments for the tuning of c_1, c_2, and c_3 parameters. The core tensor dimensions are fixed, based on the aforementioned c_1, c_2, and c_3 values.

We perform fourfold cross validation and the default size of the training set is 75 % (we pick, for each user, 75 % of his check-ins and friends randomly). The task of all three recommendation types (i.e. friend, location, activity) is to predict the friends/locations/activities of the user's 25 % remaining check-ins and friends, respectively. As performance measures we use precision and recall, which are standard in such scenarios. For a test user that receives a list of N recommended friends/locations/activities.

8.4.1 Comparison Results

In this section, we study the accuracy performance of ITR in terms of precision and recall. This reveals the robustness of ITR in attaining high recall with minimal losses in terms of precision. We examine the top-N ranked list, which is recommended to a test user, starting from the top friend/location/activity. In this situation, the recall and precision vary as we proceed with the examination of the top-N list. In Fig. 8.4, we plot a precision versus recall curve.

---

[3]http://delab.csd.auth.gr/~symeon

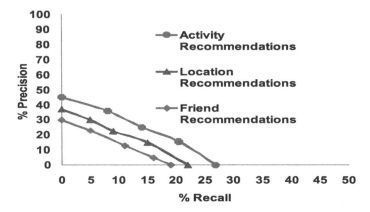

Fig. 8.4 Precision Recall diagram of ITR and FriendLink for activity, location and friend recommendations on the GeoSocialRec data set

As it can be seen, the ITR approach presents high accuracy. The reason is that we exploit altogether the information that concerns the three entities (friends, locations, and activities) and thus, we are able to provide accurate location/activity recommendations. Notice that activity recommendations are more accurate than location recommendations. A possible explanation could be the fact that the number of locations is bigger than the number of activities. That is, it is easier to predict accurately an activity than a location. Notice that for the task of friend recommendation, the performance of Friendlink is not so high. The main reason is data sparsity. In particular, the friendship network has average nodes' degree equal to 2.7 and average shortest distance between nodes 4.7, which means that the friendship network can not be consider as a "small world" network and friend recommendations can not be so accurate.

For the UCLAF data set, as shown in Fig. 8.5, the ITR algorithm attains analogous results. Notice that the recall for the activity recommendations, reaches 100 % because the total number of activities is 5. Moreover, notice that in this diagram, we do not present results for the friend recommendation task, since there is no friendship network in the corresponding UCLAF data set.

8.4.2 User Study for Location and Activity Recommendations

We conducted a survey to measure user satisfaction against two styles of explanation. The first concerns the "Peoples' Check-ins" style (denoted as style A), and the second is the "Friends' Check-ins" style (denoted as style B). For the activity recommendation, Fig. 8.6a shows the explanation style A of the GeoSocialRec[4] site, while Fig. 8.6b depicts the explanation style B.

---

[4]http://delab.csd.auth.gr/geosocialrec

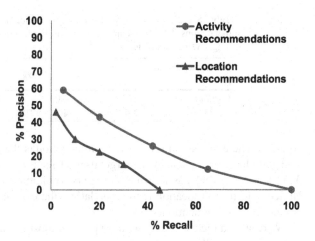

Fig. 8.5 Precision Recall diagram of ITR for activity and location recommendations on the UCLAF data set

EXPLANATION STYLE A:

We recommend the following activities based on total Check-ins!

a

Activity	Point Of Interest	POI Address	Explanation Style A: Total Check-Ins	Average Rating from style A
Sight-seeing	White Tower	Nikis Avenue--Paralia Thessalonikis	14	4.3571
Education	Aristotle University of Thessaloniki	Egnatia & Kondriktonos--Aristotle Campus	13	4.2308
Sight-seeing	Aristotelous Square	Aristotle Square--City Center	11	4.1818

EXPLANATION STYLE B:

We recommend the following activities based on the Check-Ins made by your friends!

b

Activity	Point Of Interest	POI Address	Explanation Style B: Check-Ins made by your friends	Average Rating from style B
Bar-Restaurant	Dishcotto	Analipseos 6-20, Panorama 55236, Greece--	6	3.0000
Sight-seeing	Aristotelous Square	Aristotle Square--City Center	4	3.7500
Transports	International Airport 'Makedonia' (Thessaloniki)	Kalamaria Thessaloniki--Kalamaria	4	2.7500

Fig. 8.6 Explaining recommendations based on (**a**) total peoples' check-ins, and (**b**) target user's friends' check-ins

Figure 8.6a depicts three recommended activities (Sightseeing, Education, Sightseeing) based on the explanation style *A*. As shown in the first row of Fig. 8.6a, the first recommended activity to the target user is "sightseeing" to the monument of White Tower (the first and the second column). The explanation for this recommendation is the fact that White Tower has been visited by 14 different people and got an average rating of 4.3571 in [0–5] rating scale, as shown in the last two columns of the first row in Fig. 8.6a.

Figure 8.6b depicts also a top-3 (Bar-Restaurant, Sightseeing, Transports) list of recommended activities. As shown in Fig. 8.6b, the first recommended activity to the target user is eating to a bar-restaurant named Dishcotto (the first and the second

Table 8.1 Results of the user survey for location/activity recommendations. For the mean values of the Explanation Styles, the
bigger values are better

Recommendation type	μ_A	σ_A	μ_B	σ_B	μ_d	σ_d
Location	3.77	1.13	**4.03**	1.40	0.26	0.32
Activity	3.63	0.96	**4.17**	1.44	0.54	0.31

column). The explanation for this recommendation is the fact that six check-ins in
Dishcotto have been made by the target user's friends and it got an average rating of
3 in [0–5] rating scale, as shown in the last two columns of the first row in Fig. 8.6b.
Notice that, for the location recommendation, the explanation styles A and B are
similar to the aforementioned ones.

We designed the user study with 50 pre- and post-graduate students of Aristotle
University, who filled out an on-line survey. The survey was conducted as follows:
Firstly, we asked each target user to provide the system with ratings and comments
for at least five point of interests (POIs), so that a decent recommendation along
with some meaningful explanations could be provided by our system. Secondly,
we asked them to rate separately, from 1 (dislike) to 5 (like), each recommended
location/activity list based on the two different styles of explanations. In other
words, we asked target users to rate separately each explanation style to explicitly
express their actual preference among the two styles.

We assume that, explanation style B will be the users' favorite choice, since it
relies on their friends' check-ins. Notice that according to homophily theory [3]
(i.e., "love of the same") individuals tend to prefer the same things that similar other
users do like.

Our results are illustrated in Table 8.1. The second and third columns contain for
explanation style A, the mean μ_A and standard deviation σ_A of the ratings provided
by users for location and activity recommendations, respectively. As shown, the
mean value of ratings μ_A for location recommendation is 3.77, whereas μ_A for
activity recommendation is 3.63. The fact that the mean of ratings is higher than 2.5
in the [0–5] rating scale means that the quality of recommendations is good. The
fourth and fifth columns contain for explanation style B, the mean μ_B and standard
deviation σ_B of the ratings provided by users. As shown, the mean value of ratings
μ_B for location recommendation is 4.03, whereas μ_B for activity recommendation
is 4.17. This is a clear support of the assumption that explanation style B is the
users' favorite choice.

Moreover, we computed the distribution of the difference between means of
explanation styles A and B, to verify that it is statistically significant. That is, the
difference between ratings of style A and B should not be centered around 0. Thus,
we measured the mean μ_d and standard deviation σ_d of the differences between
ratings of explanation style A and ratings of explanation style B. These values,
for each recommendation type, are presented in the sixth and seventh columns of
Table 8.1. We run paired t-tests with the null hypothesis $H_0(\mu_d = 0)$ for the two
recommendation types (i.e. location and activity). We found that for both location

Fig. 8.7 Mean and standard deviation of users' ratings evaluating explanation styles A and B for (**a**) location recommendation, and (**b**) activity recommendation

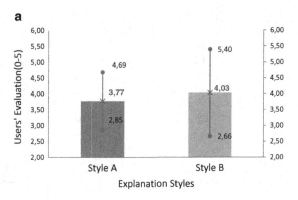

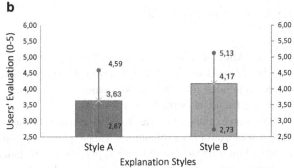

and activity recommendations, $H_0(\mu_d = 0)$ is rejected at the 0.05 significance level. This verifies the assumption that explanation style B is the users' favorite choice. Finally, Fig. 8.7a, b show a visual representation of the mean and standard deviation of users' ratings, evaluating the explanation styles A and B for both location and activity recommendation, respectively. As expected, style B outperforms A in both recommendation types (i.e. location and activity recommendation). That is, likes of our friends have a greater impact in our own choices.

8.4.3 User Study for Friend Recommendations

We conducted a second survey to measure user satisfaction against the explanation styles in friend recommendation. We have also tested two styles of explanation. Explanation style A justifies friend recommendations based on the number of common friends between the target user and his candidate friends. That is, explanation style A considers only pathways of maximum length 2 between a target user and his candidate friends. Explanation style B can provide more robust explanations, by presenting as explanation, all human chains (i.e. pathways of more than length 2) that connect a person with his candidate friends.

For instance, an example of a social network is shown in Fig. 8.8. The explanation style A for recommending new friends to a target user U_1 is as follows: "People

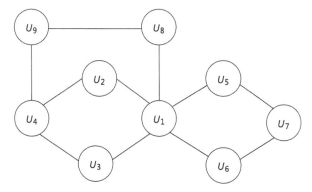

Fig. 8.8 Social network example

you may know: (1) user U_7 because you have two common friends (user U_5 and user U_6) (2) user U_9 because you have one common friend (user U_8) ...". The list of recommended friends is ranked based on the number of common friends each candidate friend has with the target user.

Based on explanation style B, a user can also get, along with a friend recommendation, a more robust explanation. This explanation contains all human chains that connect him with the recommended person. For instance, in our running example, U_1 would get as explanation for recommending to him U_4 the following human chains that connect them:

1. $U_1 \rightarrow U_2 \rightarrow U_4$
2. $U_1 \rightarrow U_3 \rightarrow U_4$
3. $U_1 \rightarrow U_8 \rightarrow U_9 \rightarrow U_4$

This means that U_4 is connected with U_1 with two pathways of length 2 and 1 pathway of length 3. It is obvious that explanation style B is analytic and informative.

We designed the same user study with the one described in Sect. 8.4.2. We assumed that explanation style B will be the users' favorite one, because it is more transparent and informative than explanation style A. Our results are illustrated in Table 8.2. As shown, the mean value of ratings μ_B of style B is 4.10, whereas μ_A is 3.8. This is a first indication supporting our assumption that explanation style B is the users' favorite choice. Moreover, we measured the mean μ_d and standard deviation σ_d of the differences between means of explanation style A and ratings of explanation style B. These values are presented in the sixth and seventh columns of Table 8.2. We run paired t-tests with the null hypothesis $H_0(\mu_d = 0)$. We found that the null hypothesis is rejected at the 0.05 significance level. This verifies our assumption that explanation style B is the users' favorite choice.

Finally, Fig. 8.9 shows a visual representation of the mean and standard deviation of users' ratings, evaluating the explanation styles A and B for friend recommenda-

Table 8.2 Results of the user survey for friend recommendations. For the mean values of the Explanation Styles, the bigger values are better

Recommendation type	μ_A	σ_A	μ_B	σ_B	μ_d	σ_d
Friend	3.8	1.13	**4.10**	0.99	0.30	0.47

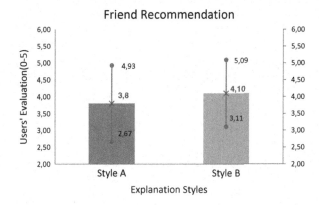

Fig. 8.9 Mean and standard deviation of users' ratings evaluating explanation styles A and B for friend recommendation

tion. As expected, style B outperforms A. That is, explanation style B increases the acceptance of a recommender system, since users can understand the strengths and limitations of the recommendation process.

References

1. M. Brand, Incremental singular value decomposition of uncertain data with missing values, in *Proceedings of the 7th European Conference on Computer Vision (ECCV)*, Copenhagen (2002), pp. 707–720
2. R. Burke, Hybrid recommender systems: survey and experiments. User Model. User-adapt. Interact. **12**(4), 331–370 (2002)
3. P.R. Monge, N. Contractor, *Theories of Communication Networks* (Oxford University Press, Oxford, 2003)
4. A. Papadimitriou, P. Symeonidis, Y. Manolopoulos, Friendlink: link prediction in social networks via bounded local path traversal, in *Proceedings of the 3rd Conference on Computational Aspects of Social Networks (CASON)*, Salamanca (2011), pp. 66–71
5. B. Sarwar, G. Karypis, J. Konstan, J. Riedl, Incremental singular value decomposition algorithms for highly scalable recommender systems, in *Proceedings 5th International Conference on Computer and Information Technology (ICCIT)*, Dhaka (2002), pp. 27–28
6. V. Zheng, B. Cao, Y. Zheng, X. Xie, Q. Yang, Collaborative filtering meets mobile recommendation: a user-centered approach, in *Proceedings of the 24th AAAI Conference on Artificial Intelligence (AAAI)*, Atlanta, GA (2010), pp. 236–241

Chapter 9
Conclusions

In this chapter, we conclude the book with a summary, and future research directions.

9.1 Summary

Chapter 2 presented fundamental information for algorithms in Recommender Systems. In particular, we have shown basic algorithms and metrics that are used in the field of Recommender Systems. Chapter 3 provided basic concepts and recommendation algorithms for Online Social Networks (OSNs). Chapter 4 focused in recommendation algorithms for the Location-based Social Networks (LBSNs). Chapter 5 presented the challenges of recommendation algorithms in LBSNs and the main algorithmic families. Chapter 6 presented state-of-the-art algorithms in LBSNs, and deepened in the algorithmic side of each method. Chapter 7 compared and categorized the algorithms that are described in Chap. 6, by taking into account their basic characteristics that differentiate them. We provided categorizations of existing algorithms in regard to the type of recommendation they provide, their data sources and data structures, the techniques they are based on, etc. In Chap. 8, we presented an example of a real-world recommender system for LBSNs.

9.2 Future Research Directions

In this section, we propose future research directions to improve the quality of recommendations in LBSNs. Firstly, we notice that almost none of the methods presented in this book has used the same experimental protocol, data sets and metrics to compare with other works. There is a lack of a benchmark evaluation method to measure the quality of the recommendations in LBSNs. Thus, an

P. Symeonidis et al., *Recommender Systems for Location-based Social Networks*,
SpringerBriefs in Electrical and Computer Engineering,
DOI 10.1007/978-1-4939-0286-6_9, © The Author(s) 2014

important future work for the whole research community in LBSNs would be the establishment of a common experimental protocol with common metrics and data sets.

Facebook has recently incorporated a new feature, namely "Suggested Guests", that returns a list of recommended people, that a target user might want to consider inviting to an event. However, it does not provide any explanation along with the recommended guests. Explaining recommendation in LBSNs will increase the systems' transparency and will increase the users' acceptance, since users can understand the strengths and limitations of the recommendation process. Future models for LBSNs should also try to provide explanations along with their recommendations. Moreover, although there are many approaches for recommending events to users, there has not been enough research work to tackle the problem of recommending people to social events. Furthermore, the study of time dimension in event recommendation might be a great challenge for the researchers of the field, over the upcoming years.

Targeted Advertising, cross-selling and social marketing in LBSNs, which can become a market of billion dollars, have not been adequately investigated from the research community. For example, an interesting advertising problem involves selecting products/services to advertise at a social network based on the current location of the users. This problem involves also Collaborative Filtering, where we try to find customers with similar behavior in order to suggest they buy products/services that similar customers have bought. Ads and product recommendation should attract more attention of the research community in LBSNs.

The exploitation of semantic meanings from additional context may also increase the quality of recommendations and user satisfaction. The semantic awareness will solve potential problems of polysemy among users' activities and locations. That is, users may have different interests for the same location, or tags/activities may have different meanings for different users.

We live in the era of "Big Data". All future models should take into consideration their scalability. Future work should go into stream mining models or data-intensive distributed applications that run on large clusters of computers (i.e. Hadoop framework) to increase their scalability. The Hadoop framework can transparently provide both reliability and data-intensiveness to LBSNs applications.

Although many recommender systems have been proposed in LBSNs, there are still open questions to be addressed. For example, will LBSNs become the next "Big Thing" of the Internet industry? There are also ethical questions ahead, concerning the quality or recommendations versus the user privacy. Nowadays, users geo-location can be inferred even for people who keep their GPS signal private. The existence of geo-location data can rise undesirable side effects during the user experience and can make any LBSN vulnerable. These privacy issues should get the attention of the research community along with the development of better recommendations in LBSNs.